UNCOVERING EINSTEIN'S NEW UNIVERSE

FROM WALLAL TO GRAVITATIONAL WAVE ASTRONOMY

UNCOVERING EINSTEIN'S NEW UNIVERSE

FROM WALLAL TO GRAVITATIONAL WAVE ASTRONOMY

DAVID BLAIR **RON BURMAN** **PAUL DAVIES**

UWA PUBLISHING

First published in 2022 by
UWA Publishing
Crawley, Western Australia 6009
www.uwap.uwa.edu.au

UWAP is an imprint of UWA Publishing
a division of The University of Western Australia

THE UNIVERSITY OF
WESTERN
AUSTRALIA

ISBN: 978-1-76080-230-1

A catalogue record for this book is available from the National Library of Australia

Cover design by Carl Knox, Australian Research Council Centre of Excellence for Gravitational Wave Discovery
Typeset in Avenir Next by Lasertype
Printed by Scotts

 uwapublishing

This book is dedicated to all the research students who contributed their vision, their imagination, their creativity and their energy to the uncovering of Einstein's new universe.

CONTENTS

CONTENTS

INTRODUCTION

This book describes the adventures of many people who between them changed humankind's common understanding of our universe and the nature of physical reality. Some of them yearned to know and understand, some were adventurers, some were inventors, and most of them shared the joy of being part of a common enterprise dedicated to improving our best understanding of the truth.

The story connects extraordinary ideas with extraordinary efforts to prove those ideas.

It uncovers the power of ideas – the ideas of a dreamy kid that gave us a new theory of reality, and his happiest thought that gave us a new theory of the universe.

To reach our new understanding it was a prodigious struggle all the way – a struggle for understanding, a struggle for measurement, and a struggle between ego and scepticism.

While our story is set in a mainly white and male-dominated world, the book uncovers unacknowledged contributions of many women and Indigenous people.

We have created this book for several reasons. One is to reveal a heroic chapter in Australia's history, one of which Australians should all be proud. Another is to tell the story of how our modern understanding of the nature of physical reality, the nature of space and time and the structure of the whole universe emerged, and to tell it in a language accessible to everyone.

Science and technology have left nineteenth-century thinking behind, but our schools have not. This book is closely linked to Einstein-First, an initiative aiming to replace nineteenth-century thinking with a consistent modern description of physical reality that everyone can learn from the early years of primary school. Children, whose abilities to understand are almost always underestimated by educators, accept Einsteinian ideas with ease, but their teachers, brought up with a nineteenth-century paradigm, often find the ideas challenging. We hope this book will find its place in school libraries where it can be a resource for teachers and students alike, contributing to history, to science and to social science.

The stories told in this book are stories of discovery and human progress and stories that look forward to exciting future discoveries. The story of Einstein, a passionate human being and the greatest genius the world has ever known, is worth telling. It is one that everyone should know.

This book was made possible by many people and organisations who all recognise the importance of science and human discovery as the foundation on which our culture is based. The University of Western Australia, the Australian Research Council Centre of Excellence for Gravitational Wave Discovery (OzGrav), Einstein-First and the Big Questions Institute are among many enlightened organisations who have supported our efforts. A full list of supporters can be found at the end of this book.

Peter Rossdeutcher, Einstein-First Initiative, August 2022.

PREFACE

In 1920 Edwin Slosson wrote a little book called Easy Lessons in Einstein. *He began his book with a dialogue, now slightly updated.*

A Dialogue: To prevent prospective readers from buying the book under false pretences.

Two commuters, each reading a newspaper, are sitting on a train.

Newspaper Reader: Here's something queer – a whole page taken with a new discovery in physics. 'Eclipse Observations Confirm Einstein's Theory of Relativity.' Anything in your paper?

The Author: Yes, and there is a cartoon about it.

The Reader: Must be something to it then. *Reading aloud*: 'Most sensational discovery in the history of science' …. 'Greatest achievement of the human intellect' …. 'Upsets Galileo, Newton, and Euclid' … 'Revolution in philosophy and theology'…It looks as though I ought to know something about this, doesn't it?

The Author: You might as well do it now.

The Reader: 'Parallel lines meet'… 'a person moving with the speed of light never grows old'… 'gravitation is due to a warp in space' … 'the length of a measuring stick depends upon the direction of its motion'… 'energy has mass' … 'time is the fourth dimension' – Why, the man is crazy, isn't he?

The Author: Well, definitions of insanity are so uncertain that it is not safe to say who is crazy. But there must be method in his madness – otherwise how could he have hit upon his exact predictions?

The Reader: Can you tell me in plain language what it's all about?

The Author: Einstein says that there are only twelve men in the world capable of understanding.

The Authors of this book: Well, Einstein may have been a genius but that sounds foolish! Anyone can understand anything if it's explained at the right level. Sometimes understanding means just getting used to a new reality. The concepts are what count and everyone deserves the right to share our best understanding of the universe.

One of the best ways of understanding is by learning from stories. This book tells some of the stories that took us from old ideas of a clockwork universe, to a new universe ruled by laws of chance, that is dynamic, evolving and never exactly predictable.

In the new universe, planet Earth is a negligible speck in a vast emptiness and simultaneously the single supremely precious place in the universe, where laws of chance created lifeforms capable of knowing and understanding its whole being, and determining our planet's future.

Modern understanding is an outcome of education and discovery by a vast assortment of scientists across the world, but behind almost all our modern science and technology we find concepts of physical reality originally discovered by Einstein.

Part 1

AUSTRALIA TESTS EINSTEIN

By David Blair and Ron Burman

Chapter 1

Albert Einstein's 'most splendid work'

I have just completed the most splendid work of my life.

Albert Einstein to his son Hans Albert, 1915

In 1915, Albert Einstein told his son, and later the world, about his theory of gravity called general relativity. It had not had an easy gestation. War was raging in Europe, millions were dying, and Einstein, a pacifist in Berlin, was doing what he could for the cause of peace.

Einstein was born on 14 March 1879, a day that today is celebrated as Pi Day, the International Day of Mathematics, because the first digits of pi are 3.14. The young Albert was a dreamy kid who was always wondering. At school they called him 'the dopey one'.

Einstein graduated with a degree in physics in 1901 and worked in the Swiss Patent Office from 1902 to 1909. He was still always wondering, and his job gave him time to think.

Then, in one miraculous year, 1905, he published four revolutionary papers that changed our way of thinking about light, matter, motion and energy.

In the first of these papers, published on 9 June 1905, Einstein proved that light comes in tiny lumps of energy that today we call photons. His paper explained the physics of turning light into electricity, called the photoelectric effect, which today is used in our cameras and solar panels.

Einstein's second paper, published on 18 July 1905, explained a strange phenomenon first observed by the botanist Robert Brown, who had visited Australia from 1801 to 1805 on Matthew Flinders' famous expedition. He was a very keen microscopist, with great interest in pollen. In Australia, Brown had described and named almost 1,200 species of West Australian plants.

Through the microscope he had noticed that tiny particles associated with pollen grains jiggled around almost as if they were alive. He published a paper describing this phenomenon in 1827, and it soon came to be known as Brownian motion. What caused this motion? Were the particles really alive?

Einstein, the dopey one, at the age of about 15.
(WIKIPEDIA COMMONS)

Brownian motion remained a mystery until 1905. Einstein explained it as motion caused by random jiggling atoms colliding unevenly with the tiny particles from the pollen. It confirmed that matter was made of atoms and molecules – a controversial idea that had long been suspected but not proved.

Einstein's third paper, published on 26 September 1905, was his famous theory of relativity that was founded on the idea that the speed of light was the speed limit of the universe. The weird thing was that the speed you measure for a light beam is always the same whatever your own speed. His theory said that as things travel faster, their mass increases, their dimensions change and time slows down. Everything in the universe was relative – except for the speed of light.

Albert Einstein (1879-1955) German-Swiss
mathematician: Relativity: Einstein in 1905 aged 26.

His fourth paper, published on 21 November 1905, introduced his famous equation $E = mc^2$, which tells us that energy has mass and when energy is released things get lighter. It tells us that our phones are lighter when their batteries are flat! When heavy atoms break up into smaller atoms there is some mass left over: this spare mass became energy in the explosion of atom bombs. $E = mc^2$ is all around us, from our phones to gas flames to the power of the Sun.

These four papers were utterly revolutionary at the time, but like most new ideas, they seem less extraordinary when we get used to them. But the 1905 papers were only the start of Einstein's revolution.

From 1907, Einstein had begun to grapple with the theory of gravity, but he was also dealing with domestic problems. He was a demanding person to live with and his 1903 marriage to classmate Mileva was in trouble. They had two children, to whom he was never close. By 1914 Einstein and Mileva were living apart and he was in a relationship with his cousin Elsa.

But Einstein was not a lonely genius. He was in constant discussion and correspondence with students and his two closest

Einstein and his first wife Mileva early in their marriage.
COURTESY ETH-BIBLIOTHEK ZÜRICH, BILDARCHIV / FOTOGRAF: LANGHANS, JAN F. / PORTR_03106 / PUBLIC DOMAIN MARK

collaborators, mathematician and college friend Marcel Grossmann and Michele Besso, an engineer in the Patent Office. Grossmann went on to help Einstein create the mathematics needed for his new theory of gravity, and Besso remained a lifelong friend.

Einstein left the patent office in 1909, and moved successively to professorships in Zurich, Prague and Berlin before becoming director of the Kaiser Wilhelm Institute of Physics in Berlin until he fled to the USA in 1933.

From 1907 onwards, Einstein struggled to solve the problem of gravity, while at the same time his marriage was collapsing and

Marcel Grossmann, Albert Einstein, Gustav Geissler and Marcel's brother Eugen during their time as students at ETH Zurich.
COURTESY THE ALBERT EINSTEIN ARCHIVES, THE JEWISH NATIONAL & UNIVERSITY LIBRARY, THE HEBREW UNIVERSITY OF JERUSALEM, ISRAEL.

Europe was marching towards war. In 1914 he moved to Berlin, his wife left him, and he and Grossmann presented the first description of gravity being connected to the geometry of curved space. Their paper was in two parts: the physics described by Einstein, and the maths described by Grossmann. Eventually, in November 1915, Einstein presented his masterwork, his full theory of gravity called general relativity to the Prussian Academy of Sciences.

Einstein's theory was later described as 'one of the greatest – perhaps the greatest – achievement in the history of human thought'. This book is about the testing of general relativity, and about its consequences. It has now passed a century of testing. It still inspires, baffles and stretches our minds, even though most of us use it every day because, without it, our GPS navigators would not work.

For Einstein, life without music was inconceivable, even as he grappled with his theory. He never travelled without his violin, 'Lina'. Einstein commented, 'I have my daydreams in music. I see my life in terms of music'.
COURTESY KEYSTONE PRESS / ALAMY STOCK PHOTO

The theory of gravity was born in the middle of World War I. While developing his theory, Einstein was actively fighting for peace. With a small group of other scientists, he published the *Manifesto to the Europeans* (see overleaf) that called for a just end to the war. Their manifesto had no immediate consequence, but the European Union we see today seems to be a fulfilment of their thinking.

After the war, one of the authors of the manifesto, Nicolai, was imprisoned by the Germans, but he escaped and eventually chose exile in Argentina. He became a leading figure in the international pacifist movement led by Britain's leading philosopher, Bertrand Russell. Much later, in 1955, Einstein and Russell drew up a manifesto against nuclear war, which Einstein signed a few days before his sudden death.

Einstein in his study in Berlin. (PUBLIC DOMAIN)

Manifesto to the Europeans, October 1914

Georg Friedrich Nicolai, Albert Einstein, Otto Buek and Wilhelm Foerster

Through technology the world has become smaller...

The struggle raging today will likely produce no victor; it will probably leave only the vanquished behind. Therefore, it seems not only good, but rather bitterly necessary, that intellectuals of all nations marshal their influence such that—whatever the still uncertain end of the war may be—the terms of peace shall not become the cause of future wars.

The time has come when Europe must act as one in order to protect her soil, her inhabitants, and her culture.

It is necessary...that Europeans get together...and then we shall try to call together a union of Europeans.

Physicists were studying Einstein's new theory even before it was published. Within months, Karl Schwarzschild had found a solution that predicted black holes. A few months later Einstein found that his theory predicted a new type of wave that he called *gravitational waves*. He predicted that these ripples of space would travel at light speed, but he thought they could never be detected and later he even doubted their reality.

From 1915 onwards, Einstein was pursuing the consequences of his discoveries. He was keen for them to be tested, and he was making predictions about photons and discovering new phenomena in quantum physics as well as making predictions about the universe as a whole. Part 1 of this book tells the story of the testing of his general theory of relativity at Wallal in Australia. This experiment convinced scientists that his theory was correct.

While the testing of his theory was underway, Einstein was becoming more and more doubtful that the universe follows the laws of chance, which was the unavoidable consequence of his discovery of photons. Part 2 of this book describes the scientific struggle to understand and prove all of Einstein's predictions, including the 99-year struggle to understand gravitational waves and how they were doubted, proved and finally detected. Part 3 is about the future of gravitational wave technology and gravitational wave astronomy.

To better understand Einstein's discoveries, we need to go back in time to the previous revolution, caused by Newton, because this revolution determined the physical reality that most adults absorbed in their schooling. Einstein overthrew this reality and gave us a completely different way of thinking about the universe.

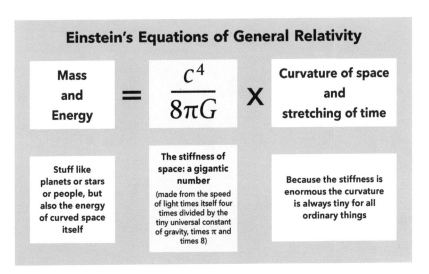

This is Einstein's theory of gravity written mostly in words. The stiffness of space is just a really huge number, so large that for modest objects like planet Earth, the curvature is tiny. If you put in your mass on the left side of the equation, you can calculate how much *you* curve the space around you. Because the stiffness of space is crazily enormous – it has 43 zeros, and is a trillion trillion trillion times stiffer than steel – the curvature caused by 50–100 kilograms must be crazily tiny!

Chapter 2

Origins of Einstein's masterpiece

Matter tells spacetime how to curve,
Spacetime tells matter how to move.

John Archibald Wheeler

Albert Einstein giving a talk in 1921.

Isaac Newton and his book, the *Principia*, where he laid out the laws of the universe.

In 1687, Isaac Newton's *Principia Mathematica* revolutionised our way of thinking. He discovered that the universe was ruled by universal mathematical laws – his famous laws of motion and gravitation.

For Newton, space was an unchangeable universal framework like a set of imaginary gridlines, time was perfectly regular, and gravity was a mysterious force of attraction between masses. Matter moved in this framework without affecting it in any way, and gravity acted according to a simple mathematical law.

But Newton could not explain one central mystery of gravity – one that had been recognised by Galileo a century earlier and noticed by several others before that. Why did wood and iron, and large and small cannonballs, all fall at the same speed? Centuries later astronauts on the moon showed the same thing using a hammer and a feather. Newton's law of gravitation gave a beautiful mathematical description of gravity but Newton himself thought that the idea of forces acting through empty space was an *'absurdity'*. Why should there be such a perfect balance between force and mass so that everything falls at exactly the same rate, no matter what it is made of?

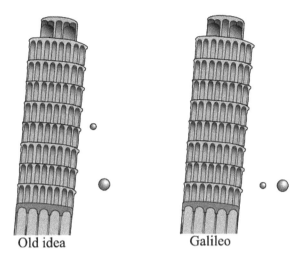

Old idea   Galileo

The ancient Greeks believed that things fall at a
speed dependent on their weight even though simple
observations show it is not true. Einstein pointed out that
if you were riding on the big weight and looked only at
the small weight, you would 'have the right to think you
were at rest'.

Galileo is reputed to have dropped objects from the Leaning Tower of Pisa
to discover the universality of freefall. Today students try this out with water
balloons at the Gravity Discovery Centre's Leaning Tower of Gingin.

Einstein started his revolution in 1905, but the world had to wait ten years for him to solve the mystery of gravity. The first clue for Einstein was in his third paper of 1905, which proclaimed that nothing can travel faster than light speed. The speed limit of the universe meant that there was no way any information could travel faster than light. But Newton's law of gravitation says that gravity arrives instantaneously. Either Einstein was wrong or Newton was wrong. This was the beginning of Newton's downfall.

In those days the first cars were very popular. Young people like Einstein must have loved to have joyrides and to feel the push-back when a car accelerates and the need to hold yourself back when the brakes go on. Perhaps it was this sort of experience that triggered Einstein's happiest thought. Here is what he said:

'When I was sitting in a chair at the Patent Office in Bern in 1907, the happiest thought of my life came to me: The gravitational field has only a relative value… for an observer falling from the roof of a house in free-fall, there is – at least in his immediate vicinity – no gravitational field at all. If, moreover, this observer drops bodies, they remain in a state of rest or uniform movement in relation to him, regardless of their physical or chemical nature (ignoring, of course, the air resistance). This observer therefore has the right to consider himself at rest.' (Einstein, 1920 cited in Spagnou, 2017)

Einstein's idea was that gravity is no different from that feeling you have in an accelerating car, except that the car's acceleration is usually horizontal and the gravitational acceleration is vertical. You feel your weight horizontally from the back of the seat pushing you forward when the car accelerates, and you feel your weight on your feet because the Earth is holding you up against the acceleration of gravity.

Einstein turned Newton's idea of gravity on its head. Suddenly he reversed gravity: it is not a *pull-down* force but a *push-up* force! It's the force that has to be exerted to prevent your natural motion. In freefall there is no force acting at all…so it is no wonder that everything falls at exactly the same rate.

This was Einstein's happiest thought. It was the mark of a genius. But even for a genius it was an eight-year struggle to turn that idea into a rigorous mathematical theory of gravity.

Finally, at the end of 1915, with the help of his friend Marcel Grossmann, Einstein had the equations and a complete description of his new theory. This was a revolution because it changed our conception of space itself.

For Newton, space was like an empty stage in a theatre, where objects like actors play out all their motions and interactions. Einstein's theory replaced Newton's plain flat stage with one that changes its shape in response to all the actors on the stage. Not only do the actors change the space around them, they also change time. Life on Einstein's stage is more like living in a bouncy castle!

Einstein's theory says that space is like a stretchy fabric: mass changes the shape of the fabric. A four-dimensional fabric is hard to imagine, but simplifying our picture to two dimensions makes it easy. The curved fabric in the image represents changes in the shape of space, and also in the speed of time.

Children learning about Einstein's theory using a stretchy lycra sheet, which is our toy version of the four-dimensional fabric of spacetime.

Light and matter all follow the shortest paths in a curved and flexible fabric. Suppose there was a steep mountain obstructing your journey across a flat plane. If you were sensible you would skirt around the lower slopes rather than climbing up to the summit and down the other side, because this path would be shorter. Every freely moving object does that in curved spacetime. Gravity is a force you feel when something like the solid surface of a planet stops you from taking the shortest path.

In Einstein's theory, the Earth takes the shortest path through the curved space and time around the Sun. A stone thrown upwards also takes the shortest path in space and time. To stop something from falling you must apply a force, and this slows down time. It takes a bit more thinking to catch this idea.

Journeys in space and time

Einstein's law of motion is often called the Principle of Maximal Aging: freely falling objects age fastest. The theory says that applying any force, like the Earth applies to us when it stops us falling to its centre, slows down time and we age slower.

Most of us take many journeys in space as we travel around our planet, but all of us take *much longer* journeys in time. What do we mean by a journey in time? Our destination in time is old age. The shortest way to that destination is to be in freefall. High above the Earth, falling freely is what you would be doing if you were in orbit like the James Webb telescope, and here aging happens fractionally faster.

Another perspective is to think about Newton's famous falling apple – or anything else that falls. An apple falls from a tree because falling is the quickest way to become an old apple. But because the Earth prevents us from falling freely, it is a time machine. It slows down time so we on Earth age a tiny bit slower.

Atomic clocks today can easily measure the slowing of time. If we lived on the surface of a neutron star where gravity is intense, a century would last about 110 years, the rest of the universe would seem to be running about 10% faster and yet your wristwatch would tell you nothing had changed!

Why did we say that our journeys in time are much longer? It is because the speed of light connects space and time. Four years of time is four *light-years* of time-travel distance – the distance light travels in four years, which happens to be the distance to Alpha Centauri, the nearest bright star. In a lifetime, your combined journeys may circle the Earth few times, but your journey in time might have a length of 80 light-years (i.e. your lifetime!), which is quite an interstellar journey!

Do you see what that means? We are all travelling in time at light speed – 300 million metres per second – while most of our travels in space are at speeds of only a few tens of metres per second.

Edwin Slosson's *Spacetime*

A line contains an infinite number of points,
A square contains an infinite number of lines.
A cube contains an infinite number of plane squares.
A tesseract contains an infinite number of solid cubes.
Its fourth dimension is time.

Spacetime is four dimensional, consisting of three dimensions of space and one dimension of time. But spacetime only makes sense if we measure space and time in the same units. By multiplying time by the speed of light, we convert the unit of time from seconds into metres: one second is three hundred million metres. That is the only way we can talk about spacetime. It took me about thirty billion metres to compose this sentence!

The principle of maximal aging emphasises that gravity is mainly due to time differences with height. On Earth the time difference is about 3 microseconds per year for every kilometre of altitude. This tiny effect on time was too small to measure until atomic clocks were invented in the 1960s, but today it is an intrinsic part of every GPS navigator. GPS navigators use space and time to determine their position on Earth. Although 3 microseconds seems like a very short time to us, it is quite a large distance in time (900 metres). It quickly makes sense that if we are to determine our place on Earth within a few metres, we must allow for the stretching of time at different altitudes.

We have already seen how the Earth is a time machine. It slows time down a little bit, although this effect is too small for us to notice. However, this gravitational slowing of time is enough that the centre of the Earth is almost 2½ years younger than the surface. Beyond our planet, the same physics – that a massive object can slow down time – tells us that time actually comes to a stop at the edge of the most extraordinary objects ever conceived: black holes. Karl Schwarzschild predicted this in 1916, thinking that it was purely abstract mathematics, but today black holes are starting to be heard almost everyday, as we will discover in Part 2.

That, in a nutshell, is Einstein's theory of gravity!

Einstein's challenge to scientists

Einstein's theory is like a magician's bag. There seems to be no end to the queer things that can be pulled out of it. The more it is studied the more paradoxical it appears.

Edwin Slosson, 1922

Einstein challenged scientists to test his predictions. The clear pre-diction that clocks run faster at high altitude could not be tested because there were no clocks accurate enough to do this. That had to wait until 1976 and that story is told in Part 2. The other possible test was to measure the shape of space.

If space is curved as shown here, light beams get extra deflection as they pass through the curved space near the Sun. Therefore a star (A) that would have been out of sight from us on Earth appears to have moved away from the Sun to position B.

In 1905 Einstein had proved that light comes as photons – tiny lumps of energy that are a bit like bullets. Einstein's famous equation, $E = mc^2$, says that energy has mass. It is just a conversion equation, like you might say an amount in cents = dollars times one hundred. Following from this equation, photons must have mass because they have energy, and so they should feel gravity. That means photons should be just like tiny bullets travelling at light speed and should be deflected by the Sun's gravity. In 1911 Einstein wrote a paper making exactly this prediction: starlight should be deflected by the Sun exactly like bullets would be if they were travelling at light speed.

A few years later Einstein realised that this was only half the answer, because his new theory said that space is curved by matter, and the place where space in the solar system is most curved must be closest to the Sun.

Curved space is a scary idea for most of us because we are in space and we can't see space, but if it is simplified to a lycra sheet it is not scary at all. Toy cars that always run straight deflect as they pass the central mass. *Straight is curved* when space is curved. Straight paths trace curved trajectories, just like aircraft route maps between distant cities. Curved space adds extra deflection to light beams – in fact Einstein calculated that it doubles the total deflection.

Unfortunately you can't see stars near the Sun, so you have to wait for a total solar eclipse of the Sun. With the Sun completely blocked out by the Moon, you can photograph stars close to the Sun. If the positions of stars during an eclipse were compared to their positions in another season when the Sun wasn't nearby, you could tell if space was curved. If Newton was right and space was unchangeable, astronomers would measure 0.87 arc seconds of deflection. If Einstein was right, they would measure 1.75 arc seconds of deflection. This would prove that space was curved near the Sun.

This tiny deflection, about 1/2000th of a degree, needed very accurate astronomy. The story of the Wallal expedition starts here.

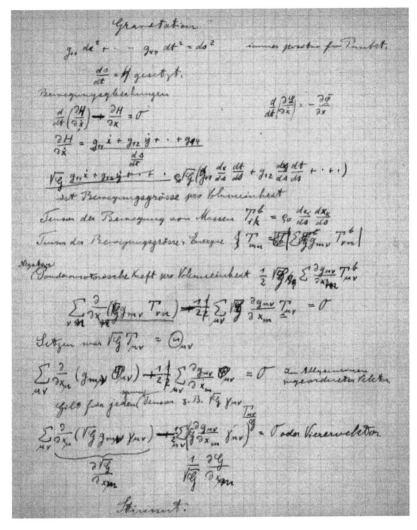

Einstein's Zurich Notebook of 1913, showing his equations linking space, time, matter and gravity. He grappled for eight years with what was to become his 1915 general theory of relativity.

Chapter 3

Testing Einstein: Fame and doubts

One of the greatest – perhaps the greatest –
of achievements in the history of human thought.

Sir J J Thomson, President of the Royal Society of London, 1919

Arthur Eddington, a British astronomer, saw the significance of Einstein's work. He rallied Britain's science community to send expeditions to observe the total solar eclipse of 29 May 1919. One expedition went to Brazil, the other to Príncipe Island off the coast of equatorial west Africa.

The expeditions did not go smoothly. In Príncipe they met torrential rain and cloud. In Brazil they had problems with telescopes and distortions caused by the solar heating of the telescope and the atmosphere.

Eddington compared the position of the stars in eclipse images with photographs of the night sky taken previously. Very few stars were visible and some images were blurred, so he made significant corrections to some of the measurements and he only measured half a dozen stars. A few stars did seem to have been deflected in line with Einstein's prediction, but some did not deflect at all and some had half the deflection predicted.

Despite these problems, on 6 November 1919 the British Astronomer Royal announced that Eddington's results disagreed with

Newton's theory of gravity and supported Einstein's predictions. This announcement made Einstein into a celebrity overnight. The front page of the New York Times read 'Lights all askew in the heavens'. Underneath it was the message: 'A book for 12 wise men. No more in all the world could comprehend it'.

LIGHTS ALL ASKEW IN THE HEAVENS

Men of Science More or Less Agog Over Results of Eclipse Observations.

EINSTEIN THEORY TRIUMPHS

Stars Not Where They Seemed or Were Calculated to be, but Nobody Need Worry.

A BOOK FOR 12 WISE MEN

No More in All the World Could Comprehend It, Said Einstein When His Daring Publishers Accepted It.

New York Times, 10 November 1919.

There were three messages: First, Newton was wrong. Second, Einstein was a genius. Third, ordinary people could not understand it!

> ## Can't Understand Einstein
> - *New York Times* headline
> 29 November, 1919

Unfortunately, the legacy of that third message is still with us today. Einstein's warning to his publisher was that few people at that time were familiar with the mathematics and concepts of his theory. That did not mean that it was incomprehensible. Edwin Slosson at the time was just completing his *Easy Lessons on Einstein*, and Einstein himself wrote a letter to *The Times* of London to explain his theory soon after Eddington's results were published. But the *New York Times*' third message did the world a huge disservice because it made everyone frightened of even trying to understand Einstein's theory.

Campbell described Eddington's eclipse expedition results as follows:

Professor Eddington's plans at Príncipe were disrupted by clouds, although a few of his plates recorded four or five star images of inferior quality.

From his measurements he deduced a displacement-coefficient of 1.61 arc seconds, agreeing closely with that predicted by Einstein, of 1.75 arc seconds, on the basis of his generalised theory of relativity.

The two cameras used in Brazil, where fortunately the weather conditions were excellent, gave Einstein coefficients differing in amount. With the astrographic camera, a duplicate of that employed in Africa, were secured sixteen plates, recording from six to eleven stars each. The images were not in good focus, apparently because the heat of the Sun's rays… caused distortion. However, all of these plates were measured.

The *New York Times* said 'lights all askew in the heavens', but Eddington's photo of the 1919 eclipse showed that if stars were out of place, it was barely detectable. The stars are visible between the horizontal lines.

The deflections measured on an A4 image were typically 1/40th of a millimetre – about 1/10th of the width of a fine pencil line.

COURTESY EDWARD WECHNER

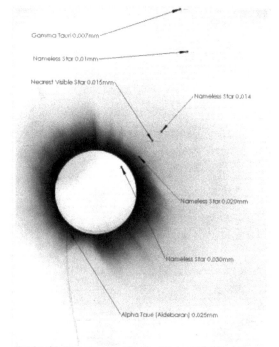

Gamma Tauri 0.007mm

Nameless Star 0.01mm

Nearest Visible Star 0.015mm

Nameless Star 0.014

Nameless Star 0.020mm

Nameless Star 0.030mm

Alpha Tauri (Aldebaran) 0.025mm

Negative of the 1919 solar eclipse taken from the report of Sir Arthur Eddington. Eddington highlighted the stars he used in the comparison with horizontal marks; these can be seen at 2 o'clock on the image.

The results for the coefficient varied from 0.00 to 1.28 arc seconds. Their mean value, 0.86, agreed well with the half Einstein effect, 0.87, predicted in 1911. The second camera gave photographs in excellent focus, and their measurement and discussion yielded an Einstein coefficient of 1.98, in good accord with the full, or double, effect 1.75, resulting from the generalized theory of relativity.

The observers attributed by far the greater weight to the last result; in fact, assigning very little weight to the first of the Brazilian results. Professor Eddington was inclined to assign considerable weight to the African determination, but, as the few images on his small number of astrographic plates were not so good as those on the astrographic plates secured in Brazil, and the results from the latter were given almost negligible weight, the logic of the situation does not seem entirely clear.

A point of interest is that the British observers were the first to say, in view of the fundamental importance of the general subject, that confirmation should be sought at the eclipse of 21 September 1922.

Alexander Ross, who we meet in the next chapter, discussed the extraordinary precision needed for the Einstein measurement:

> It is to be remembered that even the maximum possible deflection of 1.75 seconds of arc corresponds to a displacement of only I-2000th of an inch [about 13 millionths of a metre] on the plates of 5ft. [1.6 m] cameras, or to 3-2000ths of an inch [40 millionths of a metre] in the case of 15ft. [5.6 m] cameras. As the actual measurements have to be carried out on stars at some distance from the Sun's edge, the actual displacements of the observed stars range from about half down to less than one-tenth of the above amounts. These figures illustrate the extraordinary accuracy needed in the investigations.

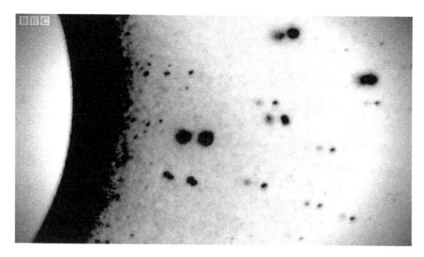

What Eddington's team would have hoped to see when they superimposed two photos, a reference photo from when the Sun was not there, and the second taken during an eclipse. The stars (the black dots in this negative image) should appear to have moved away from the Sun. But the deflections were smaller than the star images themselves and some were deflected in the wrong direction.
IMAGE COURTESY OF THE BBC

Chapter 4

The battle for Wallal

William Campbell who was the world's greatest eclipse astrono-mer was clearly suspicious of Eddington's claims, and many others shared his suspicion. Campbell's analysis quoted on page 27 shows why they were skeptical. The 1919 eclipse results certainly do not stand up to modern standards of proof. The observers could only see a few stars, and some stars actually appeared to have shifted in the wrong direction. It did not help that Einstein was German, and Germany had lost the war. More data were needed to find out if Einstein was really right!

The next total solar eclipse after 1919 was predicted for 21 Sep-tember 1922. In March 1920, the Royal Astronomical Society pub-lished the path of this eclipse. It would begin on the east coast of Africa, cross the Indian Ocean, passing the Maldives and Christmas Island, reach north-west Australia in the early afternoon, and cross the continent diagonally.

Alexander Ross of the University of Western Australia was quick to recognise the scientific opportunity of this eclipse. The eclipse path would cross the coast at Wallal, an isolated site almost

2000 kilometres from Perth. According to Ross, it was 'the most favourable site on earth' to further test Einstein's theory. But it would be no easy matter, and he immediately met opposition.

Alexander Ross

Alexander David Ross (1883–1966) was born in Glasgow and attended the University of Glasgow (MSc, 1906; DSc, 1910), which had been made famous by the iconic mathematician-physicist William Thompson, who became Lord Kelvin. Ross was at Glasgow from 1902 to 1912, during which time one of his positions was as a Thompson Fellow. By the time of his 1912 appointment as foundation professor of mathematics and physics to the new University of Western Australia, he had published about thirty papers.

Ross must have been adventurous to move from the grand Victorian architecture of Glasgow to a brand-new university with minimal resources and housed in tin sheds. He concentrated on undergraduate teaching and developing community interest in the university and in science. He was a widely travelled and popular public lecturer and broadcaster. His students respected his teaching skill, and his dapper style, vitality and wit, but out of earshot they imitated his Scottish accent.

John Budge, the last surviving member of the physics department to have known him, describes Ross as 'a dapper little Scotsman with a killer instinct — lady-killer that is! To mere men he seemed rather ordinary, but...women were drawn to him like bees to honey. Five-feet nothing tall, slim, with balding off-white hair, the waistcoat and bowtie wearing

Professor spoke his Scottish brogue in a kind of cross between a lisp and a falsetto whistle, giving the impression that he was whistling softly to himself rather than speaking…all brought about by "a crook set of snappers", according to his wife!'

Ross was an able administrator, a member of the university senate (1922–39), vice-chancellor (1918), dean of arts and science, president of the Australian Institute of Physics, president of the Royal Society of Western Australia and Western Australian secretary of the Australian and New Zealand Association for the Advancement of Science, and a fellow of the Royal Society of Arts, who awarded him a silver medal in 1951. He was appointed CBE in 1949. Ross loved music and served as president of the Perth Symphony Orchestra.

Ross met his wife Euphemia at university, where she was a student of physics and geology. They were married in the shed that comprised the University of Western Australia's physics department on 7 August 1913, a few days after Euphemia arrived at Fremantle from Glasgow.

Euphemia shared Ross's love of music. In the early days she lectured for him when he was absent and assisted with his talks in country towns. Sometimes the halls were so small that she had to operate a magic lantern from outside, projecting slides through a window. Euphemia was a pioneer of the local kindergarten movement, a foundation member of the women's college fund committee and, later, a member of the Women's College Council. She became assistant state commissioner of the Girl Guides' Association.

Ross was one of the founders of the Presbyterian Ladies College in Perth and served on the school council for 41 years. PLC has a school house named in his honour.

Wallal is around 300 kilometres south of Broome on the Eighty Mile Beach. It means 'sweet water' in the Nyangumarta language because of the fresh water source located there. The soak was identified on maps in 1879, and a government well was sunk. It became a vital watering point on the Kimberley–De Grey stock route.

The waters off Wallal were known for numerous sharks and severe storms and cyclones. They were also lucrative pearling grounds.

Wallal is on the lands of the Nyangumarta people. Their lands stretch from Eighty Mile Beach deep into the Great Sandy Desert. In 2015, the Nyangumarta Traditional Owners entered into an agreement with the Australian Government for the purposes of promoting biodiversity and cultural resource conservation. A ceremony proclaimed the establishment of the Nyangumarta Indigenous Protected Area. Nyaparu Rose, CEO of the Nyangumarta Warrarn Aboriginal Corporation stated, 'As a Nyangumarta elder, I am very proud of this achievement for the Nyangumarta People. Having our IPA will allow us to maintain our significant sites, and look after our Country – from the desert to the sea.'

Broome–Wallal signpost.
COURTESY OF THE STATE LIBRARY OF WESTERN AUSTRALIA 112585PD

Wallal became better known as a post and telegraph station on the overland telegraph line connecting Perth to the north west. In 1901 Surveyor Alfred Canning surveyed the line for the rabbit-proof fence, designed to protect Western Australia from the onslaught of feral rabbits from the east, and nominated Wallal as the fence's end point. The Wallal Downs pastoral station was established around 5 kilometres from the telegraph station in 1900.

'An isolated and lonely outpost'
Daisy Bates, 1922

By 1922 the little settlement of Wallal comprised one building for four or five operators and linesmen, and along a low ridge were the huts and shelters of more than 100 Nyangumarta people.

Telegraph linesman's house, Wallal.
COURTESY OF THE STATE LIBRARY OF WESTERN AUSTRALIA, 4131B/3/26

A hopeless part of the coast

A. R. Hinks of the Royal Geographical Society, a former astronomer, was asked by the Astronomer Royal, Sir Frank Dyson, to investigate the geographical conditions of the 1922 eclipse. He addressed the March 1920 meeting of the Royal Astronomical Society on his findings, which were reported in *The Observatory* magazine. It included the following:

'The eclipse track reaches Australia at Ninety-Mile Beach, a hopeless part of the coast, and strikes into the great desert. There are no facilities for landing. You can approach within 4 miles of the shore in small boats. The desert is inaccessible, except to camels. There are no railways within hundreds of miles, and motor cars are out of the question. The first place which is feasible is Cunnamulla in South Queensland...'

The 'hopeless part of the coast', now called Eighty Mile Beach, features tides of 8–9 metres. This enormous tide height was certainly a challenge, but Alexander Ross was not to be deterred. He and a colleague, Thomson, published a rebuttal of Hinks's rejection of Wallal as an observation site. They sent advance copies of their paper to leading observatories and eclipse committees. In particular, Ross corresponded with William Campbell, director of the Lick Observatory in California.

Ross and Thomson's paper, which was published in the *Monthly Notices of the Royal Astronomical Society*, reveals Ross's enthusiastic advocacy for Wallal. Their rebuttal of Hinks was detailed and authoritative and was instrumental in convincing the Crocker Foundation, who sponsored the Lick Observatory Wallal Expedition, that Wallal was a viable location. Ross and Thomson's paper reveals how they rebutted Hinks and persuaded Campbell.

'The solar eclipse of 1922 September 21 offers a favourable opportunity for re-testing the Einstein Relativity Theory. In view of the partial failure at Sobral in 1919 it seems highly desirable that a fair number of stations should be occupied, and we wish to draw attention to the facilities offered on the north-west coast of Western Australia.'

They pointed out that the locations under consideration for observing the eclipse, from the Maldives to Queensland, had eclipse durations of 3 to 4 minutes. The duration of an eclipse is important because it gives astronomers more time to take more photos, especially if it is partially cloudy.

'In our opinion it is highly advisable that a party should occupy a station near to Wollal, where about 5m 18s of totality is obtainable.'

Then they set about rebutting Hinks's 'hopeless' claim.

'Speaking at a meeting of the Society in March of last year, Mr. Hinks described this part of the Western Australian coast as hopeless, and stated that motor cars were out of the question. We consider that Wollal is approachable both by sea and land, although the former method would undoubtedly be preferable for the transport of instruments.'

Next, they presented information that showed they had done extensive research about accessing the Wallal site.

'Luggers of 12 to 14 tons can make the 200-mile run from Broome to Wollal in from 24 hours to 5 days, depending on the weather, which is usually calm at that period of the year. Schooners of 30 to 40 tons are also obtainable, and there is a launch of about 14 tons which could make the journey within the day…

'Whale-boats can beach with the tide at a point opposite Wollal, whence the way lies for half a mile across sand dunes and then for a mile along a rough track… There seems no reason why packages of instruments weighing two or three hundredweight should not be safely handled.

'Wollal possesses an ample supply of good water, and there is a sheep station only four miles distant from which assistance in connection with the removal and erection of instruments could probably be obtained. One resident in the district possesses a motor car, and states that the road between Broome and Wollal is fair in September although rather loose in parts.'

Ross and Thomson then addressed the practical issues of provisions for an expedition.

'If Wollal is to be occupied, the main stores should be obtained in Perth and taken north to Broome by the north-west coast and Singapore steamers. Minor stores could be purchased at Broome.

'One of the pearling fleet-owners has evinced a keen interest in the matter, and we are sure that we can obtain his assistance in making local arrangements at Broome. It is, however, imperative that early arrangements should be made for the coastal boats required at Broome and for the necessary carting at Wollal.'

They also addressed the question of the weather at Wallal.

'From the meteorological point of view Wollal gives great promise. Meteorological observations are made at 8 a.m. at the telegraph station, and the following summary will be of interest.'

They provided a table of weather data showing that over 4 years it had only rained twice at Wallal in September, and that the number of cloudy days in September varied from zero to two over the years 1917–20. Another table of data from Broome showed a similar pattern over a longer period.

Ross and Thomson concluded as follows:

'It therefore occurs to us that Wollal has many advantages in respect of weather conditions and of long duration of totality, and it is our opinion that the difficulties of transportation of instruments are not unreasonably great. We shall be glad to obtain any further information desired by any party contemplating the occupation of this district.'

Those intent on testing Einstein's theory of general relativity saw the 1922 solar eclipse as an ideal opportunity. William Campbell was convinced:

'…there seems only one way to achieve success. Know first what you want, and then go after it.'

Campbell was chairman of the American Astronomical Association Eclipse Committee. As an eclipse astronomer, he had had adventures across the world, leading expeditions to India in 1898, the USA in 1900, Spain in 1905, Flint Island, Kiribati, in 1908,

Russia in 1914, and to the USA again in 1918. His wife Elizabeth had participated 'as a volunteer' in all his expeditions from 1908 onwards.

Campbell used the information provided by Ross and Thomson to canvas support for an expedition to Wallal. The knowledge that the September climate would be mostly dry, the atmosphere would be clear and the eclipse would have a longer duration than any he had previously observed made Wallal a good choice. The 5 minute and 20 second duration would be longer than all but one of the eclipses observed since 1889.

Campbell decided to 'go after it'. He would lead his seventh eclipse expedition.

By the end of 1920, Campbell's preparations for Wallal were well underway. With limited personnel and limited funds, he invited Ross and other Australian astronomers to join him. He also offered assistance to other Australian observers planning to go to other locations and used his contacts to seek further support.

> *At Wallal, Western Australia, where the eclipse path strikes the north-west coast there will be the happy combination of the best astronomical and meteorological conditions; the longest duration of totality and greatest altitude of the sun coinciding with the clearest skies and the lowest rainfall.*
>
> H A Hunt, Government Meteorologist, 1922

In 1922 the Australian Government published a booklet about the eclipse, which included a map of the eclipse path and warnings about the dangers of viewing it with the naked eye.

Left: Australian Government booklet about the eclipse;
Below: Charting the eclipse path.
H A HUNT, THE TOTAL ECLIPSE OF THE SUN, 1922

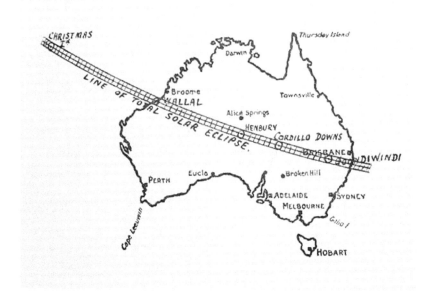

Australian Government assists

In the interests of and for the credit of Australia, it is considered desirable that every possible assistance should be given by the Commonwealth government.

John Dumaresq, Commodore of the Australian Navy,
to the Prime Minister, 1921

In June 1921 Father Edward Pigot, director of Sydney's Riverview College Observatory, wrote to Australian naval authorities on behalf of William Campbell asking for a suitable ship and help of 'a party of blue jackets' to transport Campbell and others to Wallal for the 1922 solar eclipse. Pigot highlighted the advantages of Wallal, but also its isolation and lack of accessibility.

The Australian Navy assured the prime minister that a vessel could be made available to transport men and gear. Cabinet approved the request. A group of ten naval officers under Lieutenant Commander Harold Leopold Quick were to be vital to the expedition's success.

Prime Minister Hughes designated H. A. Hunt, Esq., Government Meteorologist, as the Commonwealth official charged with the duty of conducting all further negotiations and with the general administration of the subject.

W.W. Campbell in a speech to the Astronomical Society of the Pacific

FT/SH. **PRIME MINISTER.**

SOLAR ECLIPSE IN SEPTEMBER, 1922.

Rear-Admiral J.F.Dumaresq has brought to the notice of the Acting Prime Minister a letter received by him from Dr.Pigot of the Riverview College Observatory, Sydney.

Dr.Pigot points out that the approaching total solar eclipse in September 1922 will focus much attention from the leading astronomers of the world upon Australia, for the reason that the narrow band, less than 100 miles in width, on the earth's surface where alone the total phase of the eclipse is visible, passes on this occasion right across Australia, from the North-west coast to the southern limit of Queensland.

Two of the most eminent astronomers of the world are sending out expeditions to take observations of this eclipse and have written to Dr.Pigot on the subject. They wish to locate their instruments at Wallal on the North-West Coast of Australia, where far better astronomical and meteorlogical conditions will obtain than in Eastern Australia, where the Sun's altitude at the time of the eclipse will be much lower and the length of "totality" less. Their one great difficulty is to obtain the means of transport from, say, Melbourne or Fremantle to Wallal, and Dr.Pigot inquires whether the Navy could assist by providing a suitable vessel to transport the personnel and gear to Wallal, leaving a party of Blue Jackets to assist them in the installation and picking them up a few weeks later.

In 1911 the "Encounter" took a party of six astronomers from Sydney to Tonga Island for the total eclipse of April of that year, and rendered great assistance. Dr.Pigot states that if something similar (on a smaller scale, of course) can be carried out by the AustralianNavy in connection with this eclipse, a boon will be conferred which will be appreciated by the entire astronomical world. An early decision will be necessary, as careful and elaborate preparations have to be made for the expedition.

Rear-Admiral Dumaresq, in passing on Dr.Pigot's letter, states that he regards the work as being in the highest interests of science and of advantage to Australia. He mentions that the assistance desired is such as the Royal Navy would have rendered had not the Royal Australian Navy taken its place on this station.

Mr.Hunt, Commonwealth Meteorologist, to whom the matter was referred states that Wallal is the most suitable place at which to make the observations. There are alternative methods of reaching there, e.g.,

(1) by schooner from Broome or Port Headland: a guarantee of £80 is required. Unloading is most difficult in rough weather as there is no jetty at Wallal.

(2) by motor from Broome (200 miles) or Port Headland (150 miles), but it is doubtful if cars are available for the transport of heavy instruments. One estimate is £57/10/- for the hire of a car from Broome to Wallal.

(3)/

'Two of the most eminent astronomers of the world are sending out expeditions.' Acting Prime Minister Cook notes in freehand the cabinet decision to allow the navy to assist in transporting the expeditions to Wallal.

-3-

(3) by **Lighthouse Supply Ship** "Governor Musgrave". This vessel if not at the time wanted for other work could pick up a limited supply of stores and instruments at Port Headland and land them at Wallal. The accommodation is very limited, and four months' notice is necessary to adjust coal supplies.

(4) by a suitable vessel provided by the Navy.

There is no housing accommodation or stores available at Wallal, and therefore, tents and all supplies (except mutton) must be transported.

Mr.Hunt states that the occasion is unique and the observations likely to be of much scientific value. In the interests of and for the credit of Australia, it is considered desirable that every possible assistance should be given by the Commonwealth Government.

25th July, 1921.

Cabinet, having been informed that a boat can be made available by the Navy, decides to make it available for the purpose.

Hook

Note Concerning the Total Solar Eclipse of Next Year.—

The best conditions for observing the total solar eclipse of September 21, 1922, will probably exist at or near the telegraph-telephone station of Wallal, on the northwest coast of Australia. The altitude of the eclipsed sun will be 58 degrees, and the duration of totality, about 5 minutes 20 seconds, will be longer than at any other land area within the shadow path. The region is almost rainless in September, and the sky correspondingly cloudless. The prevailing winds are described as light. The apparent objections have related to the difficulties of transport to and of landing at Wallal, and of getting away from the observing station. Wallal is about 200 miles south-southwest from Broome, the nearest point of local steamer call. Settled habitations in the vicinity appear to consist of the government telegraph-telephone station of Wallal and of the Wallal Downs sheep station, three miles southwesterly from the telegraph station. There is no landing pier. The rise and fall of the spring tide is about 18 feet. The ordinary method of reaching the station would be by fortnightly steamer from Perth-Fremantle to Broome, thence by private arrangement on small schooner from Broome to an off-shore point near Wallal, and by surf boat from schooner to shore. This procedure is more complicated and perhaps more expensive than the usual eclipse expedition follows, but the evident astronomical advantages seemed to justify it if no better method could be found.

Certain of the Australian astronomers had expressed the hope that they could prevail upon their Government to supply transport by the Commonwealth Navy. Recent cablegrams from Father Pigot, Director of the Riverview College Observatory at Sydney, and from "Weather, Melbourne," have brought the welcome news, on the authority of Sir Joseph Cook, Acting Prime Minister of the Commonwealth, that the Australian Navy would transport the Lick Observatory-Crocker Eclipse Expedition from Fremantle to Wallal. It is probable that this generous proposal is meant to apply to all seriously-intentioned eclipse expeditions desiring to locate at Wallal.

W. W. Campbell,

Mount Hamilton, California   Chairman, Eclipse Committee,
November 7, 1921.   American Astronomical Society.

REPORT OF THE ECLIPSE COMMITTEE.
By W. W. CAMPBELL, *Chairman.*

Landing one mile from the Wollal telegraph station either by beaching the boat with the tide or by surf boat is reported by reliable authorities to be entirely dependable. The telegraph operator who has lived at Wollal many years, nautical officials residing at Broome, and Australian officials elsewhere are in agreement on these points. Packages should not exceed 200 or 300 pounds in weight. A fair road leads from the beach to the telegraph and post station. A sheep ranch is located three miles southerly from the post office. There is a Government well 150 yards from the post office, with good water in abundance, and trees in whose midst camp could be established. Fuel is plentiful. Natives are available to assist in rough work. There are no mechanics in the vicinity. All supplies and expert labor should be brought from Perth and Broome, though fresh mutton in case of need can be secured at the ranch. The country back of the telegraph station is desert, practically level, with stunted vegetable growth.

The Lick Observatory definitely plans to dispatch a modest expedition to Wollal. The principal observations on the program relate to - the Einstein problem. On account of the difficulties of transport to and from Wollal, and the discomforts of long residence at Wollal, it is planned that the night photographs of the eclipse group of stars shall be secured en route at Tahiti in June, 1922, with the cameras adjusted as closely as possible to the conditions which are expected to prevail at Wollal in September. The latitudes of Tahiti and Wollal are nearly the same. In addition to the critical eclipse field of stars it is intended to photograph another selected night group of stars at both Tahiti and Wollal.

The members of the expeditions and the astronomical instruments and supplies will be transported under government auspices by commercial steamer from Freemantle to Broome, a village of 3,000 inhabitants situated about 200 miles north of Wallal. Transfer will be made at Broome to a sailing schooner for the trip to the vicinity of Wallal. The landing on the beach will be made by experienced men. The return trip to Fremantle will be by the same methods.

The Commonwealth Government has prepared to offer very extensive assistance and hospitality to the expeditions during their stay at Wallal. The Commissioners of the Commonwealth Railways have arranged for the granting of favors to the members of the eclipse expeditions for the trip from Sydney to Fremantle and return.

The generosity and hospitality of the various departments of the Commonwealth Government are on a scale quite beyond expectations, and they set a new standard.

W. W. CAMPBELL.

The Eclipse Committee of the American Astronomical Society reported in their Proceedings.

Wallal, where many of the world's astronomers will gather to view the eclipse of the sun . . . is one of the most isolated telegraph repeating stations in the Commonwealth.

Geraldton Guardian, **14 September 1922**

The first section of the eclipse expedition from the Lick Observatory, in charge of Dr. Robert J. Trumpler, reached the island of Tahiti on April 10th as planned A letter from Dr. Trumpler states that a splendid observing site was found in the garden of an American resident, with valued advantages or work rooms and other conveniences immediately at hand.

The Einstein camera of the Toronto University is scheduled to reach Tahiti the middle of May, in order that Dr. Trumpler may mount it on the Einstein polar axis of the Crocker expedition and secure the photographs of the critical eclipse group of stars, for comparison with photographs secured at Wallal on the northwest coast of Australia during the total phase of the eclipse. Professor Chant has been unable to accompany his instrument, owing to engagements which keep him at Toronto until June.

W. W. CAMPBELL.

Who were the Sun watchers?

If the astronomical expeditions bring home confirmation of the eclipse of 1919 we may have to get used to all sorts of queer ideas, beside crooked beams of light in empty space...We may get to talking about the curvature of time, the weight of heat, kinks in space, atoms of energy, four dimensions, world lines and a finite universe. We may be called upon to come to...conceive of arrows that shrink and bullets that get heavier the faster they travel; of clocks that go slower the faster they travel, and of a future that turns back and tangles itself up in the present.

Edwin Slosson, 1922

Seven international teams set off for four locations, all hoping to catch the upcoming eclipse, and all hoping for much better results than those of Eddington.

A British expedition from Greenwich Observatory went to Christmas Island to try to improve on their 1919 results. A joint Dutch–German expedition joined them.

Many teams from observatories and universities across the world headed for Wallal. The Lick Observatory team, led by William Campbell, included his assistant Robert Trumpler; Dr C Adams, the New Zealand Government astronomer; J. Hosking, a Melbourne Observatory member; and Professor Alexander Ross, foundation professor of mathematics and physics at the University of Western Australia. Perth Observatory sent another team, led by director Harold Curlewis. The Toronto University team was led by Professor Clarence A Chant. The Indian team comprised astronomers John and Mary Evershed of Kodaikanal Observatory in South India, supported by Mr Don Everson of the University of Western Australia.

In addition to the scientific teams, a photographer and film-maker, Ernest Brandon-Cremer, had been hired with a plan to document the expedition. Two self-funded British amateur astronomers also joined the teams. They were supported by the Australian Navy team under Lieutenant Commander Quick. This made the Wallal expedition a group of more than thirty.

The Wallal expedition could not have been accomplished without the support of the traditional owners of the land. The Nyangumarta

It was only through the strong and enthusiastic advocacy of Professor A.D. Ross that the expedition to Wallal was undertaken and five teams from observatories and universities around the word participated. The result was that the most accurate measurements made to that date were achieved.

John L Robins Physics Department History, UWA

people were their hosts and welcomed the international teams to their country. Their 'sweet water' was essential for the teams' survival. They helped unload the boats, drove the donkey teams, spread sand and branches to reduce the dust and assisted in many other ways. After the eclipse they performed a dramatic 'folkloric performance' for the astronomers. Scattered throughout the scientific reports are mentions of the help provided to the scientific teams.

Across Australia, Adelaide Observatory sent a group to Cordillo Downs in South Australia, while an expedition under the aegis of Melbourne Observatory, the Sydney Observatory, the University of Sydney and the Astronomical Society of NSW was sent to Goondiwindi near the southern border of Queensland.

Campbell assisted these Australian astronomers with their eclipse plans and equipment so that results could be compared.

Left: William Wallace Campbell, director of the Lick Observatory 1901–30. He went on to become president of the US National Academy of Sciences. He had just turned 60 at the time of the expedition. Right: Robert Trumpler, 35, was his assistant. He had completed his PhD at the University of Zurich, a few years before Einstein was there. He went on to make further significant discoveries in astronomy and today has a prize named in his honour.
COURTESY MARY LEA SHANE ARCHIVES, LICK OBSERVATORY

Alexander Ross in about 1950.

Clarence Chant, described as the father of Canadian astronomy, was the leader of the Canadian expedition. He was a few years younger than Campbell. Here we see him as he looked in 1935.

The Lick Observatory team at Wallal. From left to right: Dr Adams, Mrs Adams, Dr Moore, Mrs Campbell, Dr Campbell, Dr Trumpler, Lieut. Com. Quick, Professor Ross, Mr Hosking.

The astronomers' wives filled important roles at Wallal. Pictured here are, left to right, Mrs Campbell, Mrs Adams, Mrs Chant, Mrs Evershed. In photographs they can be seen assisting in constructing the telescope mounts and changing the photographic plates. Mrs Chant wrote a paper about her own observations of shadow bands, strange moving patterns seen briefly before totality begins.
COURTESY OF THE STATE LIBRARY OF WESTERN AUSTRALIA, 4131B/1/24

A grand tour to Wallal

> The trek to Wallal has been a scientific adventure. Almost all kinds of transport had been used before the party arrived at its destination. The earlier portion of the journey to Broome was made by rail, liner and coastal steamer.
>
> *Daily Standard*, **19 September 1922**

The expedition began with Swiss astronomer Robert Trumpler, from Lick Observatory, sailing to Tahiti with three 'Einstein' cameras weighing over 10 tons. There, over 3 months, he photographed the stars that 5 months later would be visible at Wallal when the Sun was eclipsed. He arrived in Sydney in late July 1922.

Dr Campbell and his wife arrived in Sydney by boat a few days later. From the start of their journey across Australia, accompanied by Lieutenant Commander Quick from the navy, the astronomers were treated like royalty. At Australian Government expense they travelled by train to Melbourne, then Adelaide. There they were joined by the Canadian team under Professor Chant. At each stop they were met by government and scientific representatives and entertained at state receptions. The Perth Observatory team and Professor Alexander Ross of the University of Western Australia joined the party in Perth on 16 August 1922.

The next day about 20 members of the eclipse party addressed a packed meeting of the Royal Society of Western Australia. Short 'lecturettes' were given by most of the international visitors. William Campbell with his wife, and Clarence Chant with his wife and daughter, were both imposing figures. Both Chant and Campbell had 20 to 30 years' experience as astronomers, and for them astronomy was a family business.

The media reported every detail. By the time they left for Broome on the coastal steamer SS *Charon*, Australians were well aware of the Einstein connection and the importance of the scientific expedition. The eyes of Australia and the world were on them.

Clarence Chant described his journey in letters. He travelled separately to Sydney, then Melbourne and Adelaide, where he had arranged to meet Campbell's party before proceeding on to Western Australia

'On August 15th at 11.45 we pulled in to the great mining town of Kalgoorlie. We were met by the Mayor and others and escorted to the Town Hall where a civic reception was given to us. Then we drove out to see the racecourse, of which the citizens are very proud.

'In the afternoon we drove to the adjoining town of Boulder, where there was another civic reception, a visit to their racecourse and also to the 'Golden Mile'.

We left at 5.15 and next morning reached Perth, the capital of Western Australia. At this place was another series of entertainments including a splendid luncheon at Parliament House at which the members of the Legislature and the Council were present in full force. This was perhaps the finest of all the functions.

On Sunday we drove to Fremantle...and there went aboard the Charon...and are now at Port Headland... We leave today for Broome where we transship to a schooner which will be towed to Wallal.

This letter will be carried out by airplane to Geraldton where it will be transferred to the railway.'

The scientific teams from the United States, Canada and New Zealand crossed Australia on the *Trans-Australian Express*, a train similar to that pictured above at Tarcoola, South Australia.

Well-wishers farewell the Wallal expedition team as it leaves Fremantle for Broome, 20 August 1922, on board the SS *Charon*.
COURTESY STATE LIBRARY OF WESTERN AUSTRALIA 4131B/1/1

The SS *Charon* at Geraldton en route to Broome with the scientists for Wallal on board. Other coastal stops were Carnarvon, Onslow, Point Sampson and Port Hedland.
COURTESY STATE LIBRARY OF WESTERN AUSTRALIA 4131B/1/2

Broome to Wallal

> *A quick run down the coast brought us to Wallal. At least someone said that it was Wallal. All we could see was a stretch of golden beach that disappeared to nothing to right and left while on it the great rollers crashed their way only to be broken up in a smother of white foam.*
>
> E Brandon-Cremer, United Theatres,
> cinematographer on the expedition, 1922

In Broome, the Sun watchers met up with the Indian team under Dr Evershed and his astronomer-wife Mary, who had sailed from Madras. They were assisted by Donald Everson from the University of Western Australia's Physics Department. Broome magistrate Colonel Mansbridge helped them buy timber and cement and had moulds made for concrete piers that they would use to support their telescopes at Wallal.

The party of scientists, their wives, Chant's daughter, two film-makers, a photographer, a Broome policeman and ten naval officers assembled on the wharf beside the two masted wooden schooner that was to land at Wallal. With 35 tons of equipment it was just too crowded on the *Gwendolen* so the ladies were accommodated on the steamer *Governor Musgrave,* which towed the *Gwendolen* south to Wallal, 300 kilometres away.

The schooner *Gwendolen* at Broome before departure to Wallal on 29 August 1922. The ladies travelled on the *Governor Musgrave*, a Commonwealth Government lighthouse vessel that towed the *Gwendolen* south. Towing was necessary because winds were low and they could not risk late arrival in Wallal.
COURTESY STATE LIBRARY OF WESTERN AUSTRALIA 4131B1/14

The landing

> The landing was rendered very difficult by the surf, which was heavy, and bumped the whale boat badly as it came to shore…a number of cases were swept away by the surf and were badly soaked.
>
> ***Daily Examiner*, 5 September 1922**

Gwendolen landing. COURTESY OF THE STATE LIBRARY OF WESTERN AUSTRALIA 4131/5/28

The *Gwendolen* arrived off Wallal at sunrise on 30 August. Landing on Eighty Mile Beach involved transferring everything to land using a lifeboat. A combination of the large tidal range (up to 9 metres) and surf made it even more difficult. Dr Campbell, 'the greatest living observer of eclipses', was the first off the boat with Lieutenant Commander Quick, keen to get preparations underway.

When the lifeboat grounded, they jumped into the water and waded to shore carrying the chronometer and other delicate instruments. They were greeted by a group from the Wallal Downs pastoral station, the Wallal postmaster/telegraph operator and around 'two scores of aborigines'. All were put to work carrying the precious packages through the water to shore.

Landing freight from the *Gwendolen* was a slow process over nearly three days, with the schooner anchored several kilometres from the shore. The unloading and carrying was done with the assistance of the Nyangumarta people.
COURTESY OF THE STATE LIBRARY OF WESTERN AUSTRALIA 4131B/1/31

Wallal Beach, unloading the *Gwendolen*'s lifeboats.
COURTESY STATE LIBRARY OF WESTERN AUSTRALIA 4131B/1/32

The process of unloading the schooner at Wallal. By now the tide was down and the Gwendolin was resting on the sand.
COURTESY STATE LIBRARY OF WESTERN AUSTRALIA 4131B/1/38

At low tide the Gwendolen was on the sand. Many willing Nyangumarta people helped transfer luggage to the donkey carts.

The local Nyangumarta people helped transport the expedition gear to the campsite.

One of the many willing helpers.
COURTESY OF THE STATE LIBRARY OF WESTERN AUSTRALIA 4131/5/30

The first night or two at Wallal were spent camped on the beach. Teams worked hard to unload equipment, erect tents and other facilities, and build stands and protection for telescopes, cameras and other equipment.
COURTESY STATE LIBRARY OF WESTERN AUSTRALIA 4131B/1/34

On the beach at Wallal.

A Nyangumarta man escorting a donkey team on the beach at Wallal.

OBSERVING SOLAR ECLIPSE.

◆

Scientific Party's Preparations,

WALLAL (W.A.), Friday.—The unloading of the schooner was completed early this morning. The last part of the work was delayed somewhat by the heavy surf. One or two packages became wet when being carried ashore from the whaleboats, but fortunately all of the cases that would suffer severely from immersion in salt water were brought ashore perfectly dry. The donkey teams are still busily engaged transporting the cargo from the beach to the camp, and this work will probably be finished tonight.

Constable Dewar who joined the expedition at Broome, has given valuable assistance in obtaining a good muster of aborigines to carry water and rock for the concrete piers. The camp presented an animated appearance last night. The whole party sat down to supper at a long row of tables. Marquees and numerous tents have been erected. Some of the party have got the wooden bases of their chief instruments put together.

The weather is uniformly fine, and there has never been any sign of mist or cloud. Dr Campbell expresses himself as well satisfied with the progress made and now that the cases have been distributed to their respective owners, the work should be still more rapid.

Building the camp

> The camp is anything but a picnic place…the ground is loose
> grey sand, and, with the clearing operations and the passage
> of the donkey teams, the air is filled with powdery dust,
> which finds its way into the food and everything else.
>
> **The Argus, 1 September 1922**

Campbell chose a suitable flat site for the observation station and camp, west of the government well, and surrounded by low trees and scrub. The navy team set up the camp, selecting sites for the twelve sleeping tents, two mess tents, two store tents and the cook's galley, adjacent to the four observing stations.

The navy team, assisted by the Nyungumarta people, unloaded the schooner 'manfully and efficiently'. Professor Chant reported:

'Professor Ross had telegraphed ahead orders for broken stone and it was ready for us. There was plenty of good sand near the well and we had brought cement with us. The aborigines carried the stone and sand and water and in a very few days the cement foundations were constructed. At the same time the instruments were taken from their cases, put together and got ready for erection when the foundations had hardened sufficiently. In the case of some of the Lick instruments ~chains and pulleys were required to lift them.'

The navy team assisted with the erection of instruments and their shelters, and generally took over camp operations. The Perth team used Wallal Downs station, about 3 kilometres away, as the base for their accommodation and instruments. Their instruments included a 12.5 inch (31 centimetre) reflecting telescope, which is still used for public observing at Perth Observatory.

Food and meals were provided by the navy, with 'near fine dining' offered. At 6.30 am each day, the camp was roused by a naval officer who walked amongst the tents 'shouting an old getting-up call used on board ship'. Work started after a quick breakfast. They had only three weeks to prepare for the big day.

Making ready

> *The making ready of the instruments was an arduous task,*
> *the work for everybody lasting from sunrise to darkness,*
> *with many extensions into the night for the purpose*
> *of securing star photographs needed in adjusting and*
> *testing the instruments.*

W W Campbell, October 1922

Everybody had an allotted task. From early morning until nightfall scientists could be seen 'using pick, shovel, or astronomical instruments with equal enthusiasm'. The women were fully involved and supported the scientists in a range of roles. Although the main

Donkey transport – 'slow but sure'. Once a load was on the shore, a donkey team hauled it over the sand ridge behind the beach. There the equipment was transferred to a large wagon and pulled by another team of donkeys along a 2-kilometre sandy track to the selected observation site.
COURTESY STATE LIBRARY OF WESTERN AUSTRALIA 4131B/1/40

emphasis was on testing Einstein's theory, studies of the solar corona were also to be undertaken. A variety of auxiliary devices providing temperature, pressure, humidity and wind measurements were set up and tested. Time signals came from Bordeaux to Wallal Downs via telegraph. The Perth Observatory team relayed them by radio to Campbell's team using a receiver at the top of the 40-foot camera tower.

Strong winds and feet stirred up dust that blew into everything, with the living area soon dubbed 'Dust Camp'. The instruments were located among the trees, but the living quarters were in an open area. Clarence Chant noted:

'In order to keep down the dust as much as possible the ground inside and around the shelter was covered with red sand carried to us by the aborigines Moses and Tommy – father and son. Much of the time this sand was kept moist with water brought to us in empty kerosene cans by the same useful people.'

To help stop the dust, the Nyangumarta helpers also brought in branches to cover the ground.

The dust made it very difficult to develop the photographic plates, as reported by Chant:

'The camera mounting was adjusted with great care in azimuth and altitude and the focus of the lens was determined by photographing the stars at night. The development of these test plates was done under difficult conditions. Mr. Owen, the line-repairer, offered us the use of his house, but we found it almost impossible to keep the developer and the fixer [chemical solutions needed for processing the plates] cool enough, and the gelatine films on the plates became soft and easily mutilated. Also, there was much dust in the air which settled on the plates if they were placed in a current of air to dry and yet if they were not so placed but kept in a closed space the films required a very long time to dry and would sometimes flow under the high temperature prevailing.

'As a consequence Dr. Young and I decided that we would not attempt to develop our critical plates at the camp at all, but would take them to Broome, after the eclipse, and try to secure better conditions there.'

Chant describes their daily routines and the difficulties with the flies as they set up for the eclipse.

'At about 6.30 a.m. the camp would be roused by the musical voice of Mr. Rhoades, who walked amongst the tents shouting out an old getting-up call used on board ship. We would rise and don our working clothes, and by this time the sun would be up and the flies would begin to swarm about us. They were the ordinary house-fly though not so large as those we had left at home. But they were very annoying and we continually wore nets hanging from our hats. The continued motion of the net as we were working kept the flies away. However they always disappeared at sunset and allowed us to rest comfortably during the cool nights. After an hour's work mixing cement or putting instruments together or making necessary alterations to apparatus we would have break-fast – a substantial and well-prepared meal – and then to work

Raising the Tower of Babel, camera rolling.

again! Lemonade was served in the mess-tent at 11 a.m. and lunch at 1 p.m. At 4 p.m. was the inevitable afternoon tea, and when darkness came on we would wash up for dinner at 7, and eat in peace after the flies had gone.'

The Tower of Babel and the Heavenly Twins

> The Lick Observatory camera pointed skyward like some huge cannon trained for distant bombardment.
>
> ***Daily Standard***, **19 September 1922**

The chief task was the erection of structures to hold the various cameras, and then their alignment. The cameras were just enormous versions of modern SLR cameras, which basically have a lens at the front and an image surface at the back end. The longer the telescope, the bigger the image size.

Modern cameras have electronic sensors, but in 1920 the sensing was done by a glass plate covered in a thin emulsion layer containing gelatine and a silver compound that turns black when exposed to light. A shutter protected the plate from the light until the moment of exposure. The biggest camera was the enormous Lick Observatory 40-foot (13-metre) camera. It was supported by the 'Tower of Babel', named by the naval officers because it was 'constructed in five different languages', one of which was Nyungumarta.

The tower supported the camera's long square tube, set at the correct angle to match the eclipsing Sun. It had a sophisticated lens of 15 centimetres diameter at its apex that created a sharp magnified image 40 feet (13 metres) below. The plate holder for its huge glass photographic plates was moved by a spring-powered clockwork mechanism that wound a cable on a drum to move the plate at the exact speed needed to follow the motion of the Sun as the Earth rotated.

The 40-foot camera (about 13-metres long) was supported by a vertical tower. A large lens at the apex focused the sun to the large photographic plate-holder mounted near the ground. Everything was wrapped to reduce dust. The camera was pointed to the sky location where the eclipse would take place.

More supports were needed for the 'Heavenly Twins', Lick's two 15-feet (4.5-metres) long and 5-feet (1.5-metres) long Einstein cameras. These cameras had four lenses. They were designed by Campbell and made by Lick Observatory. Each camera required a canvas shelter for shade as well as protection from strong winds and dust.

Alexander Ross described Campbell's Einstein cameras in the booklet he published after the eclipse.

'They were built to Campbell's design, with quadruplet lenses and very rigid girder construction, and their frames were covered with rubberised cloth, to make them light-tight. The polar axis was carried by strong wooden supports erected on cement piers. A clock mechanism paid out a wire to track the stars during the eclipse.

'Each camera was fitted with a long guiding telescope to ensure accurate positioning during the exposure. The cameras were sheltered in a louvred canvas house to maintain uniform temperature.'

Ross noted that the quadruplet lenses gave 'wonderful definition' over the 17-inch (43-centimetres) by 17-inch photographic plates. He wrote:

'The focal areas were exceedingly flat. In the case of the 5ft. cameras, the focus did not vary by so much as 1-20th inch over the plates. In the case of one of the 15ft. cameras, the variation was about 1-40th of an inch, and in the case of the other 15ft. camera, it was too small to be detected. As these lenses gave fields of 15 and 5 degrees diameter respectively, the excellence of the workmanship was marvellous.'

The main Canadian Einstein camera, with a focal length of 11 feet (3.3 metres), was smaller than those of the Lick Observatory, but it too required protection with a high wall and a roof that could be pulled aside for observations. Chant described the camera:

'The plates used in testing for the Einstein effect were of plate-glass, 3 /16 inch thick, specially supplied by the Eastman Kodak Co of Rochester. The camera was covered first with heavy farmer's satin and then with rubber sheeting. The lens, which was made by Brashear, had a clear aperture of 6 inches and a focal length of 11 feet, and it weighed 37.5 pounds.The camera tube weighed about 250 pounds.'

The two amateur English astronomers, dubbed 'the British Cavemen', created interest by setting up most of their instruments underground. The Indian team built three large concrete piers for their 21-foot (6.5-metre) focal-length camera and other equipment, protecting them with tents and a wooden structure.

The Perth Observatory team, a few kilometres away at the Wallal station homestead had the important role of accurately determining the geographic coordinates of the camp using radio time signals and astronomical measurements, so that the time that the eclipse would begin could be known exactly.

Their small twin cameras and 12-inch (30-centimetre) reflector were also set up to photograph the corona and investigate shadow bands, which are thin streaks of light and dark that move across the ground before and after the moment of totality, when the Sun's disk is completely blocked out.

At the end of the first week, the scientists, including Robert Trumpler, seen here 'after a dip', were able to spend a day at the beach, swimming and collecting shells. Dr Evershed spent much time catching and classifying lizards, moths and butterflies.

COURTESY STATE LIBRARY OF WESTERN AUSTRALIA, 4131B/5/17

Countdown

The weather was perfect in the lead up to the eclipse, with the clear nights needed for adjusting the equipment. The second week saw numerous rehearsals for the big day. Clarence Chant reported:

'Mr James Kean, one of the naval party, would stand on a box with a chronometer before him. A certain second would be chosen for the beginning of the total phase, the zero hour, and at precisely six minutes before that epoch Mr Kean would call out 'Six minutes before!' Everyone would assemble be at his post. Four minutes later he would call 'Two minutes before!' – Then 'thirty seconds before!' Director Campbell would be looking through the finder of his Einstein camera and when he would (in imagination) see the moon just cover the sun and the corona flash out he would shout 'Go!' Then Mr Keane would count the seconds, One! Two! Three... Fifty-nine! One Minute! One! Two! etc until some seconds after the time totality was supposed to end. This practice continued until the fateful day arrived.'

At last, the big day: 21 September 1922

THE SHADOW BANDS AT THE AUSTRALIAN ECLIPSE

By JEAN L. CHANT
Journal of the Royal Astronomical Society of Canada 1923,
Vol. 17, p. 379

There was some doubt in my mind as to whether I should see the shadow bands, but no doubt as to the wonderful view I was to have of the eclipsed sun, as I was away from the little wattle trees and absolutely alone, with nothing to distract my attention. Some considerable time after first contact, when the moon had moved well over the face of the sun, the astronomers took their places at their instruments and I stood beside one of the sheets. "Six minutes before!" called out Mr. Kean, who was. announcing time for the American and Canadian parties. The darkness was rapidly coming on and the landscape assumed a peculiar greenish hue, as if a storm were approaching. The seconds went by and I felt rather nervous as this was my first experience of the kind.

Suddenly, and before I had expected it, the shimmering, elusive wave-like shadows began to sweep over me and the sheet. I grasped the rod and moved it back and forth until it was parallel to the crests of the waves as they moved forward. At the same time I began counting seconds- "one-and, two-and, three-and," etc. -and I continued counting up to 150 before totality was announced. I think perhaps I counted a little too rapidly and, allowing for that, I judge that the bands began approximately 2 minutes before totality. They continued to move over me for only a short time, perhaps ten seconds, and then they were gone! They were faint, thrilling, ghost-like, but definite enough for me to be sure that I had seen the shadow-bands.

... I was awed and filled with wonder at the beautiful sight in the sky. As there was a little time free I made a sketch of the corona

... Having closed the camera I took my position at the second sheet to watch for the shadow-bands again. They began at 15 seconds after totality and to my great surprise they seemed to come from the opposite direction. They were even fainter than at the beginning and lasted perhaps five seconds. I moved the second rod until I judged it was parallel to the crests of the waves. At the beginning of totality the bands came from west to east, moving in the same direction as did the moon's shadow as it swept across the earth; at the end they moved in the opposite direction, or so it seemed to me.

Perth Observatory reported the eclipse to the Western Australian Astronomical Society as follows:

'The Sun rose on the 21st in a perfectly clear sky. As the day wore on, the temperature increased, reaching a maximum of 93 deg. at 12.15p.m., just after first contact…

'As totality approached, the fowls retired to their roosts, and the sheep, horses and cattle came from under the shelter of the trees and commenced to feed, just as they are accustomed to do in the cool of the evening at sunset. The flies, which were very numerous during the whole of our stay at Wallal, were much affected by the changed conditions, and appeared to become quite paralysed, so that one could pick them up as if they were dead.'

Alexander Ross gave a different perspective:

'Twenty minutes before totality there was a distinct change in the general illumination of the landscape, the light becoming of a somewhat livid color. The sky had turned a darker blue…

'Twelve minutes before totality the sky had assumed a yellowish color round the horizon…

'Four minutes before totality the light was very striking in its color, objects in the landscape appearing much as if viewed at normal times through yellowish sun glasses. Thereafter the light became much more livid, and by two minutes before totality white objects assumed bluish and purplish tints.'

The weather was perfect. At 13.40 hours, on 21 September 1922, 10 minutes before the moon's silhouette would completely cover the Sun, the signal for action sounded. After final adjustments to instruments, slides for cameras were placed in position, shutters were tested and retested, and all gathered around their allotted instruments to await totality. Several newspaper reports said that the Nyungumarta people were frightened by the sudden darkening of the Sun and kept out of sight.

The temperature dropped by about 5°C, the birds were silent, and the moment of totality arrived. The eclipse program started, and it was progressed 'with clockwork precision'. There was only

5 minutes and 15.5 seconds to obtain the evidence that would prove or disprove Einstein's theory.

New Zealand astronomer Dr Adams, assisted by his wife, took exposures from inside the Lick 40-foot camera, the navy's Mr Keane provided audio time signals and Campbell was in charge of the smaller of the Heavenly Twins, with two naval officers changing the glass plates. Trumpler guided the 15-foot cameras with assistance to operate the plate holders. Not one second was wasted.

From Ross's report, which he published the next year beside Campbell's own report, it seems clear that Ross must have decided well before the expedition that he was not going to be just another pair of hands for the Lick team. Instead, he conducted a well-planned, vigorous, wide-ranging set of observations of all the phenomena that occur during a total solar eclipse. His paper is a vivid description of the whole event. Perhaps of greatest interest are his observations of the phenomenon of shadow bands. These were also studied by Mrs Jean L. Chant, who published her own description of her observations in the *Journal of the Royal Astronomical Society of Canada* in December 1923. *She writes:*

'There was some doubt in my mind as to whether I should see the shadow bands, but no doubt as to the wonderful view I was to have of the eclipsed sun, as I was away from the little wattle trees and absolutely alone, with nothing to distract my attention.

'Some considerable time after first contact, when the moon had moved well over the face of the sun, the astronomers took their places at their instruments and I stood beside one of the sheets. 'Six minutes before!' called out Mr. Kean, who was announcing time for the American and Canadian parties. The darkness was rapidly coming on and the landscape assumed a peculiar greenish hue, as if a storm were approaching. The seconds went by and I felt rather nervous as this was my first experience of the kind.

'Suddenly, and before I had expected it, the shimmering, elusive wave-like shadows began to sweep over me and the sheet. I grasped the rod and moved it back and forth until it was parallel to the crests of the waves as they moved forward. At the same time

I began counting seconds 'one-and, two-and, three-and, etc.' and I continued counting up to 150 before totality was announced.

'I think perhaps I counted a little too rapidly and, allowing for that, I judge that the bands began approximately 2 minutes before totality. They continued to move over me for only a short time, perhaps ten seconds, and then they were gone! They were faint, thrilling, ghost-like, but definite enough for me to be sure that I had seen the shadow-bands.

'I then got ready to expose a camera which Dr. Young had prepared.'

After reporting on her photography, Jean Chant went on:

'At the beginning of totality the bands came from west to east, moving in the same direction as did the moon's shadow as it swept across the earth; at the end they moved in the opposite direction, or so it seemed to me.'

> *Two minutes before totality. The landscape assumed in turn a yellowish tinge then a greenish blue, then a purple colour, and the shadows cast by the narrow crescent sun were sharp and harsh.*
>
> **Alexander Ross, University of Western Australia**

At the end of that memorable day, Campbell, Chant and the Perth Observatory team all expressed satisfaction. Others were not so happy. The British amateurs were disappointed because their self-recording magnetic instrument had failed to register the eclipse.

The Indian team reported on their results in the *Kodaicanal Observatory Bulletin* in early 1923. They had brought 3 tons of equipment. Ross had lent them his able technician Don Everson, who worked alongside Mary Evershed in operating the instruments during the eclipse. Enormous efforts had been taken to support them. Their report was full of thanks to Ross and Everson and:

'...to the steamship companies who generously carried our three tons cargo of instruments free or at a nominal charge, and to the ships' officers who handled it with scrupulous care...[and] the Royal Australian Navy who catered for us generously and provided commodious tents...[and] Commander Quick and his men for the assistance given and for their care of our instruments, which were landed at Wallal and brought back to Broome under difficult conditions.'

Unfortunately, Dr Evershed's report was a catalogue of failure.

'The spectrograph plates were total failures...the two short exposure plates in some unexplained way had been badly fogged...The fifteen seconds exposure plates showed movement of the star images and poor definition of the corona due to the bad driving of the coelostat...This completed the failure of our eclipse expedition.'

Under ideal conditions all their photographic plates had failed 'for one reason or another....Failure under the ideal conditions of a perfectly clear sky, with excellent definition and a long duration of totality, is deplorable, especially when public funds have been risked.'[1]

Evershed also said:

'Our admiration for the American installation was perhaps tinged with envy...if British manufacturers could be induced to abandon the old methods and apply ball bearings to all moving parts in astronomical instruments, as should have been done thirty years ago, an enormous gain would result.'

But despite this failure, Evershed's enthusiasm for the expedition and for Australia was entirely positive. The people of Broome helped him a lot.

'We should like also to refer with gratitude to the welcome we received from the inhabitants of Broome generally, who gave us every assistance and took a keen interest in our work. At their request I gave a public lecture on the Sun...The Resident Magistrate of Broome, Col. Mansbridge very kindly allowed me to build a

---

1   A coelostat has a moveable mirror that can be used to reflect an image of an object, for example, into a fixed telescope. The mirror is moved to track an object across the sky so that the object does not appear to move.

pier in his compound for use on our return from the eclipse in an attempt to photograph a high dispersion spectrum of Canopus.'

It is worth remembering the extraordinary precision needed for the photographs. The tiniest bit of stretching of the gelatine-based emulsion is sufficient to distort star positions. Ross pointed out:

'...the maximum possible deflection of 1.75 seconds of arc corresponds to a displacement of only I-2000th of an inch [about 13 millionths of a metre] on the plates of the 5ft. cameras, or to 3-2000ths of an inch [40 millionths of a metre] in the case of the 15ft. camera. As the actual measurements had to be carried out on stars at some distance from the Sun's edge, the actual displacements of the observed stars ranged from about half down to less than one-tenth of the above amounts.'

Despite these risks, Campbell was confident that the photographs that they had collected would be 'of great value to science', but he would not say more. When asked about the results he responded, 'Men of science feel the necessity of absolute conviction before venturing public expression of results.'

Sun Worshipers: The movie

Throughout the expedition there had been a plan to make a movie, recording the science and the people involved, including the Indigenous people.

Ernest Brandon-Cremer, an adventurer, film-maker and newsman, had been hired to make the movie. At just 27 years of age he had already had an extraordinary life. He had been born to a theatrical family in 1895 and spent his early childhood touring Australia and New Zealand in a vaudeville troupe. By age 11, his parents' marriage had collapsed, and he had run away from home. By age 14 he had stowed away on a boat from London to Hong Kong, performed in Chinese theatres, learnt movie making in Seattle and San Francisco, and become an assistant cameraman in New York. The next year he was part of the creation of the

Hollywood movie industry. By 1922 he had worked on numerous movies in New York, Los Angeles and London, fought in World War I, filmed in Morocco with the Spanish Foreign Legion, and had just arrived back in Australia.

Many of his photographs are held in archives, but the loss of his 1-hour movie is a tragedy. Just 2 minutes survives. To our knowledge it was the first scientific documentary ever made. It was played all over the world, including at the Royal Albert Hall in London. Brandon-Cremer had struck up a friendship with fellow

Australian aviation pioneer Charles Kingsford-Smith and astronomers at Wallal. Alexander Ross's booklet about Einstein's theory includes an aerial photograph of the site attributed to himself, from which we deduce that Ross also had a joyride with Kingsford-Smith.

adventurer Charles Kingsford-Smith, who helped him take aerial photos of the expedition. Six years later Kingsford-Smith would become the first person to fly across the Pacific Ocean from the USA to Australia.

Western Australia's aviation pioneer Norman Brearley, founder of Western Australian Airways, also visited Wallal and is shown in the photo overleaf at Wallal with Kingsford-Smith. We found no record of whether these adventurous airmen were present for the eclipse. They delivered mail and offered at least one joyride.

Kingsford-Smith at Wallal with his Bristol Type 28 Coupe Tourer biplane.

Ernest Brandon-Cremer with a movie camera
at Wallal.

Charles Kingsford-Smith (left) with aviation pioneer Norman Brearley (right) at
Wallal.

Cinema Film of the Total Eclipse of the Sun at Wallal, Australia, September 21, 1922.

A wireless apparatus was erected to keep the eclipse party in communication with the outside world, and a weekly aeroplane service was instituted. The film is well worth seeing by those interested in the work of scientific expeditions. It would have been too much to expect that a film of this kind, taken under such difficult conditions, would come up to the standard of the films produced by special actors in artificial conditions. However, the fact that the actual work of the astronomers is interspersed with pictures illustrating the life of the natives should make the film one of more general interest. With these additions the showing of the film takes a little over an hour. The attempt to produce a film showing the actual work of a scientific expedition is one which deserves every encouragement and we wish it every success.

ECLIPSE FILM

-----+-----

"SUN WORSHIPERS"

_____

ASTRONOMERS PRAISE

EXCLUSIVE TO "THE DAILY MAIL"
LONDON, Saturday — An exhibition
began in the Albert Hall today of a
film entitled 'The Sun Worshipers"
which is the official record of
Campbell's international eclipse
expedition to North-West Australia.

Astronomers are loud in their
praises of the film which is the work
of a 40-foot camera. Motion pictures
are being shown of the actual eclipse
which turned the face of the sea to
lurid red. Sand dunes, bush scenes
and aboriginal corroborees form the
background of the picture which is
accompanied by a scenic record of
Australasia

Leaving Wallal

The final crucial phase in the project was yet to come: careful comparison of the star images taken in Tahiti before the eclipse with images of the same stars taken during the eclipse when their light had passed near the Sun. Hopes for an announcement in Broome were dashed. The development of the huge glass photographic plates was started at Wallal, but temperatures in the darkroom tent could not be controlled, the night air was often moist and dust was an issue.

The teams decided to take the plates to Broome, where suitable space could be found to develop them and ice could be obtained for cooling. The camp was dismantled, and the instruments and plates were carefully packed and taken by donkey team to the shore.

The departure from Wallal was delayed because of rough seas after a storm. One lifeboat was sunk by a big wave, and its contents could not be retrieved until it was exposed by the low tide the next day. Getting the luggage through the rough seas and loaded onto the schooner moored some miles offshore was very difficult. Eventually the Wallal Sun watchers returned to Broome on 28 September.

Gear being transported through the surf at Wallal. One boat was sunk by a wave.
COURTESY OF THE STATE LIBRARY OF WESTERN AUSTRALIA 4131B/5/28

The team took their treasure trove of 270 kilograms of glass photographic plates to the Broome coastal radio station. The buildings they used still stand; they are now the toilets of the Broome Bowling Club!

The developed plates revealed the very large number of stars they had recorded. This was very good news. Now they had the images required to study, measure and compare positions with the same stars that Trumpler had photographed in Tahiti. The plates were transported to Sydney, then, in November, were shipped to the United States for analysis.

All that remained of the eclipse expedition at Wallal were the cement pillars that had held the instruments and a plaque that Campbell had promised in memory of the project.

Campbell's report on the expedition given to the Astronomical Society of the Pacific in 1923 includes much fascinating detail and background, but also includes the following note:

'Many of the conditions existing immediately prior to and during the total phase, ably observed by Professor Ross, are described in his interesting paper which appears elsewhere in this number of the Publications.'

He concludes his article with extensive thanks to Australia and the many Australians who contributed to its success: the thanks give an indication of how the Wallal eclipse expedition captivated the whole country.

'A few names, however, must be given, as to leave their services unmentioned would be unpardonable. First of all, there is Mr. H. A. Hunt, Commonwealth Meterologist, Melbourne, the Government official charged by Prime Minister William M. Hughes and his Cabinet with the duty of making all arrangements, in so far as these concerned the Australian Commonwealth and the Australian States of New South Wales, Victoria, South Australia and Western Australia, through which the expedition journeyed.

'To the Prime Minister, Senator Pearce and the Cabinet, who favored us with a formal luncheon in the Parliament Building and who authorized the granting of all items of assistance considered

The plaque commemorating the Wallal expedition.

essential to the success of the expedition or contributing to the comfort of its members, grateful acknowledgments are made.

'The formal luncheon tendered to our expedition and others by the Premier of Western Australia, Sir James Mitchell, the members of his Cabinet and the members of both the houses of Parliament in the Parliament Building; the assignment of the S. S. Governor Musgrave to tow the schooner Gwendolen from Broome to

Wallal, and many other items of encouragement and helpfulness, were concrete evidences of the spirit of interest which prevailed throughout the entire State.

'The expedition is deeply indebted to the higher officials of the Royal Australian Navy, and especially to Lieutenant-Commander Quick, who remained with us constantly from the moment of our arrival at Sydney, as guide, philosopher and friend, until we returned to Melbourne on the east-bound trip; and also to Chief Warrant Officer Rhoades, R.A.N., who served as Commander Quick's principal assistant.

'The welfare of the expedition was constantly held in mind by His Excellency Sir George Murray, Lieutenant-Governor and Chief Justice of South Australia, by the Nestor of pure science in Australia, Professor Sir T. Edgeworth David of Sydney University; by Professor Alexander D. Ross of the University at Perth; by Director Curlewis of the Perth Observatory; by Mr. W. F. Gale of Sydney; by Sir Thomas Lyle, Professor Emeritus of Physics in the University of Melbourne, and by hosts of others.

'The universal interest in our expedition and its purposes, and in astronomy in general, prevailing throughout the Commonwealth, was indicated not only by the very great assistance afforded by the Commonwealth and State Governments, but also by the great number of receptions, official, and private, the more than two score of addresses and lectures delivered by my colleagues and me, by the character and number of newspaper articles conveying accurate information, always upon a substantial and dignified basis, and by the kindly interest on steamers and railways everywhere. The hospitality extended, the interest shown, and the assistance afforded, were of a standard higher than I have ever observed in any other part of the world on any occasion.'

The world waits

After the results of the photographs then taken have been measured we may perhaps know whether Einstein is to be ranked with Copernicus and Newton, among those who have revolutionized man's conception of the universe, or whether he will be regarded merely as the author of an ingenious mathematical theory of limited applicability to reality.

Edwin Slosson

The British, Dutch and German expeditions to Christmas Island all failed due to cloud. So it all depended on the Wallal results to prove Einstein's general theory of relativity. The media clamoured for news, and the scientists were under intense pressure. But Campbell firmly reiterated that there would be no hasty announcements.

In October 1922 the University of Western Australia awarded William Campbell an honorary degree of Doctor of Science as 'a foremost representative of astronomers in the world'.

Campbell did not start work on the plates for some time. The precious cargo arrived at the Lick Observatory in December 1922, but Trumpler did not return home until February 1923. It was only then that both he and Campbell could throw themselves into processing the plates.

Campbell and Trumpler independently selected and measured three pairs of Wallal plates showing from sixty to eighty stars. They then compared them with those plates taken by Trumpler in Tahiti 3 months before the eclipse. They found that the displacements of the stars close to the eclipsed Sun had a mean value of 1.74 seconds of arc. The value expressed in Einstein's theory was 1.75, so the agreement was almost exact.

Predictions sustained

We not repeat Einstein test next eclipse.

Cable from W W Campbell to Sir Frank Dyson,
Astronomer Royal, 12 April 1923

On 12 April 1923 Campbell cabled Einstein in Berlin to say that the Lick Observatory team had confirmed his prediction. A cable to the Astronomer Royal expressed Campbell's confidence that there would be no need to repeat the Einstein test at the upcoming September 1923 eclipse. Chant's Toronto team also released its results. Although not as definitive, they also confirmed Einstein's prediction.

The quest for truth was over. It had been proven at Wallal, with the support, interest, generosity and hospitality of the Australian Government, the Nyangumarta people and Australians across the whole country.

RELATIVITY.
—
WALLAL PARTY.
—
PHENOMENAL RESULTS.
—
Predictions Sustained.
—
NEW YORK, April 12

Sydney Morning Herald, 13 April 1923.

EINSTEIN THEORY VINDICATED.

By an astronomical correspondent.

Professor Albert Einstein's theory of Relativity has received within the last few months observational verification which would appear to place it on an unassailable foundation.

During the total eclipse of May 1919, several photographs were secured which seemed to indicate an appreciable deflection.

The eclipse of 1922, however, afforded the most decisive evidence. Professor Campbell, director of the Lick Observatory in California, himself not personally biased in favour of the theory, went to Australia and personally superintended the work of the Lick eclipse expedition. On April 12 last Professor Campbell announced that the photographs showed an appreciable deflection in the case of from 62 to 84 stars in the field surrounding the eclipsed sun. The mean observed value by the deflection he announced as 1 3/4 seconds of arc which is identical with the actual value of the deflection predicted theoretically by Einstein.

The *Guardian*, 4 February, 1924

TIME, SPACE, AND GRAVITATION:

Letter to the *Times* of London

By Dr. Albert Einstein

I respond with pleasure to your Correspondent's request that I should write something for the *Times* on the Theory of Relativity.

After the lamentable breach in the former international relations existing among men of science, it is with joy and gratefulness that I accept this opportunity of communication with English astronomers and physicists.

It was in accordance with the high and proud tradition of English science that English scientific men should have given their time and labor, and that English institutions should have provided the material means, to test a theory that had been completed and published in the country of their enemies, in the midst of war.

Although investigation of the influence of the solar gravitational field on rays of light is a purely objective matter, I am none the less very glad to express my personal thanks to my English colleagues in this branch of science; for without their aid I should not have obtained proof of the most vital deduction from my theory.

The theory of relativity is a theory of principle. To understand it, the principles on which it rests must be grasped. But before stating these it is necessary to point out that the theory of relativity is like a house with two separate stories, the special relativity theory and the general theory of relativity.

Since the time of the ancient Greeks it has been well known that in describing the motion of a body we must refer to another body. The motion of a railway train is described with reference to the ground, of a planet with reference to the total assemblage of visible fixed stars.

In physics the bodies to which motions are spatially referred are termed systems of coordinates. The laws of mechanics of Galileo and Newton can be formulated only by using a system of coordinates.

The state of motion of a system of coordinates cannot be chosen arbitrarily if the laws of mechanics are to hold good (it must be free from twisting and from acceleration).

The system of coordinates employed in mechanics is called an inertia-system. The state of motion of an inertia-system, so far as mechanics are concerned, is not restricted by nature to one condition. The condition in the following proposition suffices : a system of coordinates moving in the same direction and at the same rate as a system of inertia is itself a system of inertia. The special relativity theory is therefore the application of the following proposition to any natural process : *Every law of nature which holds good with respect to a coordinate system K must also hold good for any other system K', provided that K and K' are in uniform movement of translation.*

The second principle on which the special relativity theory rests is that of the constancy of the velocity of light in a vacuum. *Light in a vacuum has a definite and constant velocity*, independent of the velocity of its source. Physicists owe their confidence in this proposition to the Maxwell-Lorentz theory of electro-dynamics.

The two principles which I have mentioned have received strong experimental confirmation, but do not seem to be logically compatible. The special relativity theory achieved their logical reconciliation by making a change in kinematics, that is to say, in the doctrine of the physical laws of space and time. It became evident that a statement of the coincidence of two events could have a meaning only in connection with a system of coordinates, that the mass of bodies and the rate of movement of clocks must depend on their state of motion with regard to the coordinates.

...The older physics, including the laws of motion of Galileo and Newton, clashed with relativistic kinematics...Physics had to be modified. The most notable change was a new law of motion for (very rapidly) moving mass-points, and this soon came to be verified in the case of electrically laden particles. The most important result of the special relativity system concerned the inert mass of a material system. It became evident that the inertia of such a system must depend on its energy-content, so that we were driven to the conception that *inert mass was nothing else than latent energy.* The doctrine of the conservation of mass lost its independence and became merged in the doctrine of conservation of energy...

The application of this general theory of relativity was found to be in conflict with a well-known experiment, according to which it appeared that the weight and the inertia of a body depended on the same constants (identity of inert and heavy masses). Consider the case of a system of coordinates which is conceived as being in stable rotation relative to a system of inertia in the Newtonian sense. The forces which, relatively to this system, are centrifugal must, in the Newtonian sense, be attributed to inertia. But these centrifugal forces are, like gravitation, proportional to the mass of the bodies. Is it not, then, possible to regard the system of coordinates as at rest, and the centrifugal forces as gravitational? The interpretation seemed obvious, but classical mechanics forbade it.

This slight sketch indicates how a generalized theory of relativity must include the laws of gravitation, and actual pursuit of the conception has justified the hope.

But the way was harder than was expected, because it contradicted Euclidean geometry. In other words, the laws according to which material bodies are arranged in space do not exactly agree with the laws of space prescribed by the Euclidean geometry of solids. This is what is meant by the phrase "a warp in space."

The fundamental concepts "straight," "plane," etc., accordingly lose their exact meaning in physics.

In the generalized theory of relativity, the doctrine of space and time, kinematics, is no longer one of the absolute foundations of general physics. The geometrical states of bodies and the rates of clocks depend in the first place on their gravitational fields.

Thus the new theory of gravitation diverges widely from that of Newton...But in practical application the two agree so closely that it has been difficult to find cases in which the actual differences could be subjected to observation. As yet only the following have been suggested :

1. The distortion of the oval orbits of planets round the sun (confirmed in the case of the planet Mercury).

2. The deviation of light-rays in a gravitational field (confirmed by the English Solar Eclipse expedition).

3. The shifting of spectral lines toward the red end of the spectrum in the case of light coming to us from stars of appreciable mass (not yet confirmed).

The great attraction of the theory is its logical consistency. If any deduction from it should prove untenable, it must be given up. A modification of it seems impossible without destruction of the whole.

No one must think that Newton's great creation can be overthrown in any real sense by this or by any other theory. His clear and wide ideas will forever retain their significance as the foundation on which our modern conceptions of physics have been built.

A final comment. The description of me and my circumstances in The Times shows an amusing feat of imagination on the part of the writer. By an application of the theory of relativity to the taste of readers, today in Germany I am called a German man of science, and in England I am represented as a Swiss Jew. If I come to be regarded as a *bete noire*, the descriptions will be reversed, and I shall become a Swiss Jew for the Germans and a German man of science for the English !

The 1922 Wallal expedition proved Einstein's prediction that space is curved by matter (see page 90 for Einstein's own description of his theory). But in 1916 Einstein had made another revolutionary prediction: that space can have ripples. This prediction turned out to be even more controversial, and took far, far longer to be resolved. Australia would again play a major role in testing this, Einstein's most far-reaching prediction. Today it has given humanity a brand-new sense: an ability to listen to the universe and to probe the absolute limits of space and time. The sounds of rippling space are like ripples in the surface of a drum – the drumbeats of the universe. With these gravitational waves humans are no longer deaf to the universe, and we are embarking on a new journey of discovery as we explore this new spectrum

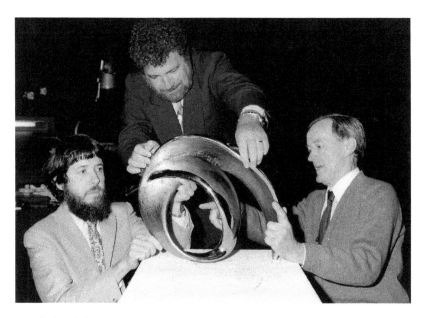

David Blair (left) with Michael Buckingham and Remo Ruffini (centre) after The University of Western Australia received the Marcel Grossmann Award for contributions to general relativity in 1988. The silver sculpture depicts orbits around a spinning black hole.

Sixty-nine years later the University of Western Australia was rewarded for Alexander Ross's efforts. The university received the Marcel Grossmann Award for contributions to understanding Einstein's general theory of relativity. By this time UWA was deeply involved in the search for gravitational waves.

Imagining ripples in space from coalescing objects. In reality the ripples are in the volume of three dimensional space and more difficult to visualize.

Part 2

STRUGGLES AND DISCOVERIES

By David Blair

Chapter 1

Black holes, gravitational waves and cosmology

The eclipse expeditions had proved a central prediction of Einstein's theory of gravity, which explained gravity as the curvature of spacetime. During an eclipse, the stars were 'askew in the heavens' – actually just shifted by the tiniest amount because the Sun, even with its immense mass, only curves space by about one part in a million.

Before he had published his full theory, Einstein had tried to solve his new equations of general relativity, but he was only able solve them approximately. Even so, he was able to show that the curved space around the Sun would change planetary orbits. He predicted that planetary orbits should be subtly different from the perfect ellipses predicted by Newton's theory of gravity. They should *precess*, so as to trace out a daisy pattern as shown in the image overleaf.

Astronomers had already observed an unexplained distortion in the orbit of Mercury. Einstein's calculation fitted with these observations, so that by the time he published his full theory, he had one piece of evidence that his theory might be correct. (described in Einstein's letter to *The Times* in Part 1).

Karl Schwarzschild was Germany's top astronomer. He was director of the Potsdam Astrophysical Observatory, but when the war started in 1914, he joined the German army. Out on the Russian front, where he was calculating the trajectories of bullets and the like, he read Einstein's paper and realised that he could solve Einstein's theory if he assumed a perfectly spherical mass.

Schwarzschild wrote up his solution and posted it to Einstein on 22 December 1915. He concluded the letter by writing: 'As you see, the war treated me kindly enough, in spite of the heavy gunfire, to allow me to get away from it all and take this walk in the land of your ideas.'

Einstein replied 'I have read your paper with the utmost interest. I had not expected that one could formulate the exact solution of the problem in such a simple way. I liked very much your mathematical

Because of the curvature of space, planetary orbits are not perfect ellipses fixed in space as Newton predicted, but precess, as shown here. Einstein proved that the very tiny distortions in Mercury's orbit that astronomers had already observed were in agreement with his theory of gravity.
IMAGE CREDIT: ESO (PUBLIC DOMAIN)

treatment of the subject. Next Thursday I shall present the work to the Academy with a few words of explanation.'

Karl Schwarzschild's solution was extraordinary. It predicted that if you kept increasing the mass in a certain volume, then there would suddenly be a point where time came to a stop, and no light could ever escape. He had discovered *Schwarzschild singularities*, which today we know as black holes. Einstein and others assumed them to be a mathematical peculiarity, because nothing like this could possibly exist in the real world. Mathematics often threw up difficulties like this.

Schwarzschild wrote:

'Often I have been unfaithful to the heavens. My interest has never been limited to things situated in space, beyond the moon, but has rather followed those threads woven between them and the darkest zones of the human soul, as it is there that the new light of science must be shone.

Only a vision of the whole, like that of a saint, a madman, or a mystic, will permit us to decipher the true organizing principles of the universe.'

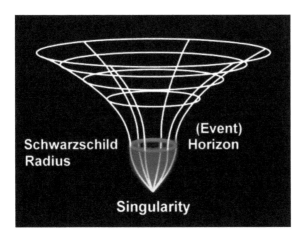

Schwarzschild's solution for a black hole. All matter is concentrated at a point called the singularity. No light can escape from inside the horizon, where it appears to a distant observer that time has come to a standstill.

In March 2016 Karl Schwarzschild returned home from the front suffering an incurable autoimmune skin disease, maybe brought on by a Russian gas attack. He died in May and Einstein spoke at his funeral:

'He fought against the problems from which others fled. He loved discovering the relations between multiple aspects of nature, but what drove his search was joy, the pleasure an artist feels, the vertigo of the visionary capable of discerning the threads that weave the fabric of the future.'

On 22 June 1916, very soon after Schwarzschild's death, Einstein presented another approximate solution to his equations. This solution was a prediction of waves. These waves were ripples in the curvature of space that he called (in German) *Gravitationswellen* or gravitational waves. These waves, he predicted, would travel at light speed, and would be emitted whenever masses accelerated, as they do when two stars orbit one another. He gave a simple formula for calculating the energy emitted in gravitational waves and commented that 'in all thinkable cases' the energy has 'a vanishing value'. He thought gravitational waves were of academic interest only.

The idea of gravitational waves makes sense to physicists today because we have got used to the idea that space is an elastic fabric, and we all know that elastic things can ripple and stretch like the surface of a drum. But in 1916 understanding depended crucially on the difficult mathematics of curved space. To make it even more difficult, Einstein had made an error in his paper, which he corrected two years later. His 1916 paper marked the beginning of 40 years of controversy before physicists could agree on the theoretical reality of these waves.

In 1917, with war still raging, Einstein decided to apply his new theory of gravity to the universe as a whole. To Dutch astronomer Willem de Sitter, he wrote 'For me…it was a burning question whether the relativity concept can be followed through to the finish or whether it leads to contradictions.'

To create a theory of the whole universe, Einstein made one crucial assumption: the universe must be static! That to him was plain common sense. Who could believe that it was not permanent and here forever?

But Einstein had a problem. His equations could not give a static solution any more than a ball in mid-trajectory could remain static and unchanging. You would need something to hold it up. This something he called the cosmological constant. It was quite an arbitrary thing to do, a bit like adding $1 million to your bank statement because you want to be rich!

Einstein with Abbe George Lemaître, who helped convince Einstein that the universe could not be static and unchanging.
COURTESY RUE DES ARCHIVES / GRANGER.

Einstein's attempt to describe the whole universe in a single equation set the scene for all modern cosmology. The first to follow was a young Russian called Alexander Friedmann, in 1922. He threw out Einstein's demand for a static universe that stays unchanging forever, and replaced it with a dynamic universe that must be either expanding or contracting. Five years later, the Belgian Jesuit George Lemaître, unaware of Friedmann's work, found the same solution, but he also showed Einstein evidence from astronomy that distant galaxies looked redder because they were moving away from us. This meant that the universe could not be static, it must be expanding. Einstein rejected all of this, saying 'this looked to him abominable!'

In 1930, Einstein went to Cambridge to visit Sir Arthur Eddington, who had made him famous with his eclipse results. Eddington had recently pointed out another problem with Einstein's static

Einstein visits Eddington in Cambridge, who points out that Einstein's solution for the universe is unstable.

universe – it was unstable. It was like balancing a broomstick on end: any perturbation would make it collapse.

Eddington may have convinced Einstein, but he refused to admit defeat until he visited Pasadena in 1931. There, at Mt Wilson Observatory, the *New York Times* reported Einstein saying: 'The redshift of distant nebulae has smashed my old construction like a hammer blow'. He swung his hand to indicate the blow. 'The only possibility is to start with a static universe lasting a while, and then becoming unstable and expansion starting, but no man would believe this.'

Soon afterwards Einstein teamed up with Willem de Sitter, and together they created a new model for an expanding universe that was completely free of the cosmological constant. The Einstein–de Sitter universe was widely accepted. Many years later Einstein told a few physicists that the cosmological constant was 'the biggest blunder of my life'.

But there is an amazing modern twist to this story of Einstein's 'blunder', because in 1998 a new discovery re-instated Einstein's cosmological constant. Two teams of astronomers discovered that the universe is not merely expanding, but its expansion is accelerating. This needed the cosmological constant to describe the force that drives this acceleration…which seems to be just the sort of instability that Eddington had seen in Einstein's equation. Whatever is causing this runaway acceleration is now called dark energy. Australian astronomer Brian Schmidt shared the Nobel Prize for this discovery. It is still the biggest mystery in our understanding of the universe.

Chapter 2

Einstein, the doubter

Science is driven by scepticism. All discoveries are provisional, and all theories are to be tested and questioned. Even though it took a few years for Einstein to admit the smashing of 'my old construction', as a true scientist he was ready to criticise his own discoveries and was not overly attached to them. In the 1930s his scepticism came to the fore.

Today we recognise that Einstein's discoveries revolutionised the way we imagine the universe. His theory of gravity revolutionised the way we think about space and time and about the whole universe. Two key consequences were the existence of black holes and gravitational waves.

Einstein's 1905 discovery that light comes as photons revolutionised our understanding of the small-scale workings of the universe. Its consequence was the theory of quantum mechanics, which tells us that, at its heart, the universe is ruled by laws of probability and uncertainty. We will see later that gravitational wave detection combines both of these crucial theories. If the three consequences – black holes, gravitational waves and quantum uncertainty – were false, the idea of gravitational wave detection would be nonsense.

Yet twenty years after announcing his theory of gravity, Einstein was busy trying to disprove *all three* of those key consequences. Two of them were errors of understanding but the third one was prophetic. It is an amazing story.

Einstein's discovery that light comes as photons was shocking not only for physicists but for society, because it completely changed the conception of how things happen. It threw out the old comfortable idea that everything that happens is directly caused by something else, and that everything in the universe runs like clockwork, precisely following the laws that Newton and others had discovered.

The simple change from light being a wave to light coming as a stream of photons immediately told physicists that, at its foundation, everything must be statistical, and that the world is ruled by chance. Why so, you may ask?

It is quite easy to see why this is. You only need to look through a window to understand it. If you see your reflection in a window, you know that you are seeing photons that reflected off the transparent glass and into your eyes, even though most of the photons hitting the glass go right on through. How could this happen? What tells the photons whether to reflect or go straight through?

A comfortable explanation that Newton would have liked would be that every single photon splits into two pieces, with a small piece being reflected and a big piece going straight through. This would be a neat deterministic process, with no need for any uncertainty. But if that did happen, the *energy* of the photons would have to be shared two ways as well. But photon energy depends on colour – blue photons have almost twice the energy of red. The split photons would have to have a dramatically different colour because of their different energy. We all know that this does not happen – a blue sweater still looks blue in a window reflection.

Every deterministic explanation for window reflections fails. But it is easy to explain if you are willing to accept probability: photons have a precise *probability* of reflecting. Partial reflections occur because there is a roughly 4% chance of photons reflecting off a window, and a 96% chance of them going through. (The exact percentages depend on the type of glass and a few other things.) The photons reflect randomly as if you were throwing a pair of dice for each of them, with photons only reflecting when you score a double six.

Einstein hated this conclusion. In 1926 he wrote to another leading physicist, Max Born, saying 'The theory produces a good deal but hardly brings us closer to the secret of the Old One…I am at all events convinced that He does not play dice.' (By, He, and 'the Old One', he meant God or the Laws of Physics.)

Einstein now set out to prove he was right. He debated and argued with other physicists, and especially with the Danish physicist Niels Bohr, who had used Einstein's discovery to craft a new model of the atom and quantum mechanics. Finally in 1935, with his two postdocs (postdoctoral research associates) Boris Podolski and Nathan Rosen, he came up with a proof that quantum mechanics predicts something that anyone in their right mind would agree was absurd nonsense.

He proved that quantum theory predicts what he called 'spooky action at a distance'. It says that if I prepare photons in pairs in a certain way, then a measurement on one of them can determine the outcome of a measurement on the other, wherever the other one happens to be, even if it is far away in the universe.

Today this paper is one of the most famous papers in quantum physics. It is called the EPR paper (after their initials), and their prediction is called the EPR paradox.

However odd it sounded, Einstein's proof was correct. Quantum physics does predict this spooky action at a distance. Einstein, Podolski and Rosen had discovered a new level of quantum weirdness called *quantum entanglement*. It was 15 years before entanglement was observed.

Einstein's attempt to disprove the statistical nature of the quantum world actually provided a brilliant proof that the quantum world was even weirder than anyone had previously imagined. It was one of the most important discoveries of quantum physics and is now a key part of quantum engineering. It is used for creating quantum computers and is used to improve gravitational wave detectors.

Einstein's doubts did not end here. Next, he came up with proof that gravitational waves do not exist! He submitted a paper to *Physical Review*, the pre-eminent physics journal in the USA, but it was heavily criticised by referees and he angrily withdrew it. This time he was wrong! His anger was only human, but his willingness to try to disprove his previous discoveries showed that he was a true scientist, ready to fault his own work. But it also showed that even he had difficulty in understanding the concept of the waves he had predicted.

In 1939 Einstein questioned the concept of Schwarzschild singularities, which we now know as black holes. In a paper in a mathematics journal, he concluded: 'The essential result of this investigation is a clear understanding as to why the "Schwarzschild singularities" do not exist in physical reality.' Again, he was wrong, but there would not be conclusive evidence for black holes until many years after his death, and even now, huge questions remain: what is inside a black hole, and is information lost forever inside a black hole?

In 1935, Einstein and Rosen, came up with an amazing concept originally called an Einstein–Rosen Bridge, but now known as a wormhole. These are curved passages between different locations in spacetime. While appearing to be theoretically possible, we do not know whether they actually exist. However, they are really useful to writers, who have used them to power many science-fiction stories. If they do exist, they would allow extraordinary spacetime journeys, as well as a way of extracting energy from nothing.

Einstein died in April 1955, leaving behind confusion – were gravitational waves real, even in theory? Could there really be Schwarzschild singularities? How might they be connected?

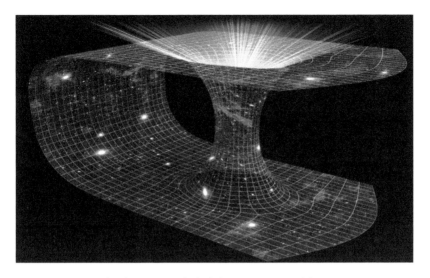

An Einstein-Rosen bridge or wormhole linking two parts of the universe.

Chapter 3

1957: Gravitational waves exist in theory

Einstein was not the only person to suggest that gravitational waves do not exist. Eddington had once famously said that 'gravitational waves travel at the speed of thought'.

Two years after Einstein's death, Richard Feynman, the bongo-drum-playing prankster and safe-cracker who won the 1965 Nobel Prize in Physics, settled this question of the reality of Einstein's waves. He did it using the powerful logic of a *thought experiment* – an imaginary scenario in which one can work through all the physics without the messy detail of a real experiment.

Feynman described two masses that he called A and B, spaced some distance apart, and gravitational waves passing them. 'Let one mass A carry a stick which runs past touching the other [mass] B. I think I can show that the second, in accelerating up and down, will rub the stick, and therefore by friction make heat.' He then went on to demonstrate this using an equation derived by Pirani. Finally he concluded 'In view therefore of the detailed analysis showing that gravity waves can generate heat...I conclude also that these waves can be generated and are in every respect real.'

What Feynman had shown was that gravitational waves make real measurable changes in the distance between objects. You can think of it in terms of inertia. Inertia is the coupling of matter to space. Usually we think of space as unchanging. Then inertia is experienced as resistance to acceleration through space. But if space expands or contracts, then the coupling of matter to space means that the matter experiences an acceleration that is provided by the expanding space. This is a real acceleration creating real changes in distance between objects.

Gravitational waves can be measured by light just like light is used to measure the speed of cars on a road. Measuring gravitational waves, in theory, is not unlike measuring the expansion of the universe. The waves are an undulating expansion and contraction in opposing directions, while the expanding universe is a uniform expansion. The reddening of light from distant galaxies proved to Einstein that the universe is expanding, and Feynman's proof meant that the colour of distant objects would fluctuate back and forth between more red and more blue.

Richard Feynman, who resolved the issue of the reality of gravitational waves, and also made enormous contributions to understanding the quantum world.

Gravitational waves, as ripples in the shape of space, are ripples of geometry. They ripple the shape of objects and they stretch and shrink distances. The rubbing stick in Feynman's thought experiment gets hot because the spacing between two masses changes.

Feynman knew that you could never detect the waves using sticks and thermometers, but his thought experiment was sufficient to prove that the waves were real because they could deliver energy.

The next step was for physicists to devise practical ways of detecting the waves. One idea was to measure the gravitational wave induced by stretching huge heavy masses. Another would be to use lasers beams to measure the distance between distant masses. But Einstein had shown that gravitational waves were vanishingly small. Could there be any waves strong enough to be detected?

Before we discover the answer to this question, we need to go back to the 1930s and meet two controversial scientists.

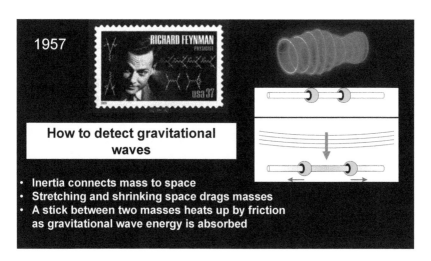

Richard Feynman, shown here on a US postage stamp, and his thought experiment where massive beads on a rod are stretched back and forth by expanding space, which is illustrated as a distorting sausage shape in space.

Chapter 4

Neutron stars and black holes imagined

Fritz Zwicky was an outspoken Bulgarian astronomer at California Institute of Technology (Caltech), a user of the world's biggest telescopes. He was fascinated by the sudden appearance of brilliantly bright new stars in distant galaxies. He called them *supernovae*. What could they be?

In 1932, the discovery of the neutron by James Chadwick had completed the puzzle about the structure of atoms: they consisted of a cloud of electrons around an extraordinarily tiny atomic nucleus composed of protons and neutrons, about 50,000 times smaller than the atom. Neutrons are slightly heavier than protons – it is as if electrons and protons are squeezed together to make neutral particles with no electric charge.

A year after Chadwick's discovery of the neutron, Zwicky came up with a dramatic explanation for supernovae. He suggested that inside huge old stars, the positively charged protons and negatively charged electrons get squeezed together to make neutrons, in which the charges have cancelled. The inner part of the star collapses into a ball of neutrons, while the outer layers explode in a

If a competition were held for the most unrecognized genius of twentieth century astronomy, the winner would surely be Fritz Zwicky (1898–1974).

IMAGE SMITHSONIAN MUSEUM

A bold and visionary scientist, Zwicky was far ahead of his time in conceiving of supernovas, neutron stars, dark matter and gravitational lenses. His innovative work in any one of these areas would have brought fame and honours to a scientist with a more conventional personality.

But Zwicky was anything but conventional. In addition to his brilliant insights that turned out to be right, he also entertained notions that were merely eccentric. To his senior colleagues he could be arrogant and abrasive. He referred contemptuously to 'the useless trash in the bulging astronomical journals.'

He once said, 'Astronomers are spherical bastards. No matter how you look at them they are just bastards.' His colleagues did not appreciate this aggressive attitude and, mainly for that reason, despite Zwicky's major contributions to astronomy, he remained virtually unknown.

TEXT COURTESY OF THE AMERICAN MUSEUM OF NATURAL HISTORY

cosmic fireworks display. The object left behind, that triggered the explosion, he called a *neutron star*.

A neutron star would be an extraordinary thing if it could exist: it would be as dense as an atomic nucleus, and although its size would be only 20 kilometres, the size of a city, its mass would be like the Sun, about one million Earths.

Robert Oppenheimer had an office down the hall from Zwicky. He, too, was a controversial figure. He was an eccentric genius and a child prodigy who had collected rocks, written poetry and studied French literature. He was abrupt, impatient and antagonised his fellow students. By the 1930s he had had a stellar career, working with the world's leading physicists, but falling out with some of

Einstein conferring with Oppenheimer, around 1950.
WIKIPEDIA PUBLIC DOMAIN.

them. He was in such demand from the top US universities of Harvard, Caltech and Berkeley that he had shared appointments, first between Harvard and Caltech, and later between Berkeley and Caltech. During World War II, he would be appointed to direct the Manhattan Project to build the first atom bomb.

In 1938 Oppenheimer decided to investigate Zwicky's idea of neutron stars with his Russian-born student George Volkoff. He had to combine the quantum physics of neutrons with Einstein's theory of gravity to find out if such a star were possible.

Oppenheimer and Volkoff's result was astonishing. They found that as you add more neutrons to a neutron star, it actually gets smaller. The greater the mass, the smaller it gets, and, at a certain mass, a bit less than the mass of the Sun, the star ceases to be stable – it just collapses inwards. This meant that there was a *critical mass* for such balls of neutrons. They concluded that 'actual stellar matter after exhaustion of thermonuclear sources of energy will, if massive enough, contract indefinitely'.

Oppenheimer pursued the collapse question with his student Hartland Snyder, concluding with an uncanny description of the formation of a black hole:

'When all thermonuclear sources of energy are exhausted a sufficiently heavy star will collapse…this contraction will continue indefinitely. The total time of collapse for an observer comoving with the stellar matter is finite…an external observer sees the star asymptotically shrinking to its gravitational radius.'

The last sentence describes a remarkable property of black holes: if you were to fall in, your own wristwatch would tell you it was happening quickly, but your friend observing from a safe distance would see the fall slowing down indefinitely, so she would see you stuck at the edge, never finally disappearing into it.

Apart from Zwicky, most astronomers dismissed the idea that neutron stars might actually exist, so Oppenheimer's results were mere academic curiosities. There were more pressing things to think about as the World War II was beginning. Besides, there was no direct evidence for neutron stars or black holes.

Einstein 1939: "*The essential result of this investigation is a clear understanding as to why Schwarzschild singularities do not exist in physical reality.*"

Oppenheimer and Volkoff 1939
"*On Massive Neutron Cores*" : critical mass

Oppenheimer and Snyder 1939: "*When all thermonuclear sources of energy are exhausted a sufficiently heavy star will collapse. ... this contraction will continue indefinitely. The total time of collapse for an observer comoving with the stellar matter is finite... an external observer sees the star asymptotically shrinking to its gravitational radius.*"

In 1939 Einstein claimed that black holes do not exist, while Oppenheimer and his students explained exactly what would happen as a black hole formed. Five years later when the first atom bomb exploded in Nevada, Oppenheimer said 'Now I am become Death, the destroyer of worlds', but his words could equally have applied to the black holes that he ignored thereafter.

After Oppenheimer, all progress in understanding Einstein's theory of gravity stopped for the next 20 years. Einstein's colleague Freeman Dyson later commented in his 2006 book *The Scientist as a Rebel* that Oppenheimer's prediction of black holes 'was by far his greatest scientific achievement'. It was fundamental to the modern development of astrophysics, 'and yet he never showed the slightest interest in following it up. So far as I can tell, he never wanted to know whether black holes actually existed...How could he have remained blind to his greatest discovery?'

Feynman's proof that gravitational waves are real happened at a famous 1957 conference in Chapel Hill, North Carolina, which took place a few months before the first artificial Earth satellite *Sputnik* 1 was launched. The conference marked the start of a resurgence in efforts to understand Einstein's general relativity. It was attended by many who would shape gravity research over the next two decades. One attendee was Michael Buckingham, who went on to become professor of theoretical physics at the University of

Western Australia and helped initiate gravitational wave research in Australia.

As the 1960s approached, the theoretical reality of gravitational waves was settled, but Schwarzschild singularities, despite Oppenheimer's work, were still a mathematical peculiarity. Curved space had been measured at Wallal, but warped time had not. A complete proof of general relativity would require much more evidence.

An experimental program to detect gravitational waves was started by Joseph Weber at the University of Maryland in the early 1960s. About the same time astronomers began to speculate about 'degenerate stars' and their possible collapse. The evocative name *black hole* first appeared in print in *Science News* in 1964, and neutron stars began to be taken seriously. An army of theoretical physicists, including John Archibald Wheeler and Freeman Dyson, who had both been at Princeton with Einstein, and Kip Thorne and Stephen Hawking, began investigating the properties of the newly named black holes, while others investigated the concept of neutron stars.

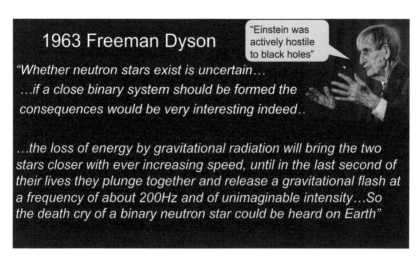

1963 Freeman Dyson

"Einstein was actively hostile to black holes"

"Whether neutron stars exist is uncertain…
…if a close binary system should be formed the consequences would be very interesting indeed..

…the loss of energy by gravitational radiation will bring the two stars closer with ever increasing speed, until in the last second of their lives they plunge together and release a gravitational flash at a frequency of about 200Hz and of unimaginable intensity…So the death cry of a binary neutron star could be heard on Earth"

Freeman Dyson, Einstein's last surviving colleague, who died in 2020, was the first person to realise that vast amounts of gravitational wave power could be created by pairs of neutron stars. [Quotes from *Interstellar communication*, New York: Benjamin Press, 1963, Chapter 12, and Dyson interview: https://www.youtube.com/watch?v=kEAy6ClrDLA].

In 1963 Freeman Dyson wrote an extraordinary article called 'Gravitational machines', in which he considered the possibility of advanced civilisations signalling to each other by making enormous bursts of gravitational waves. He did a simple calculation using Einstein's original formula for gravitational wave power, but instead of thinking about normal matter, he asked what would happen if you could arrange for two neutron stars to orbit each other:

'...the loss of energy by gravitational radiation will bring the two stars closer together with ever increasing speed, until in the last second of their lives they plunge together and release a gravitational flash at a frequency of about 200 Hz and of unimaginable intensity...The death cry of a binary neutron star could be heard on Earth if it happened once in ten million galaxies. It would seem worthwhile to maintain a watch for events of this kind, using Weber's equipment or some suitable modification of it.'

Dyson's prediction was a huge boost for Joseph Weber's pioneering gravitational wave detection program, which we will look at later. Weber himself showed me a prediction even better than Dyson's. If only he had believed the reality of Schwarzschild singularities, Einstein might have done it himself.

Einstein had calculated that two stars orbiting each other make tiny amounts of gravitational waves, basically because stars are huge and diffuse and their gravity is weak. For example, if two Suns were in orbit around each other, they would only curve space by about one part in a million, which was the curvature observed during the Wallal solar eclipse. If only Einstein had imagined two black holes orbiting each other! Then he would have found a very different answer.

In 1973, Weber showed me a few lines of high-school-level algebra that Einstein could have done within minutes of reading Schwarzschild's solution in 1916. It showed that the gravitational wave power for two black holes in close orbit with each other is given by a number that in symbols is c^5/G, which means the speed of light multiplied by itself five times, divided by the tiny constant of gravitation. This is one of the biggest numbers known in physics (more than ten to the power of 52, or one with 52 zeros!) It represents a power greater than that of all the stars in the universe combined!

Unfortunately, Einstein never did this calculation because he did not believe black holes could be real. If he had, he would have seen that gravitational waves were not merely of academic interest but of immense importance to the energy budget of the universe.

Weber's calculation showed that pairs of black holes are gravitational machines that convert mass and energy into the energy of rippling space. This energy spreads throughout the universe at the speed of light like a cosmic tsunami. Matter is turned into rippling spacetime – not all of it, but about 5% of the mass of a black hole pair. The rest of it becomes a new bigger black hole. Weber's algebra provided optimism that Einstein's waves might really be detectable.

Chapter 5

Joseph Weber: The maverick pioneer of gravitational wave astronomy

For most physicists gravitational waves were in the too-hard basket. They agreed with Einstein's initial observation that they were of academic interest only. It needed a maverick, an independent thinker, to go against the prevailing opinion. Joseph Weber was just such a maverick. He had been a US naval officer and had a distinguished war record including the sinking of a Japanese aircraft carrier in the Battle of the Coral Sea.

In 1950 Weber invented the concept of the laser, a device that would use a quantum phenomenon called *stimulated emission* that had been predicted by Einstein 33 years earlier. The word LASER, standing for Light Amplification by the *Stimulated Emission* of Radiation, was not Weber's word but he was the first to recognise the practical possibilities offered by Einstein's idea.

Weber shared his ideas with Charles Townes, a professor at Columbia University. Townes used Weber's results and went on to share the 1964 Nobel Prize for the physics of the laser with two Russian physicists. Sixty years later we use Weber's idea and Einstein's stimulated emission every time we scan a barcode at a supermarket checkout!

Joseph Weber, the Naval officer. WIKICOMMONS

Weber was justifiably bitter about not winning the Nobel Prize, and remained secretive for the rest of his life. He set himself a new Nobel Prize–worthy ambition of discovering Einstein's gravitational waves. He began to build the first gravitational wave detectors in the 1960s. His idea was to search for and see if any cosmic sources of gravitational waves might exist. Given the vanishing smallness of gravitational waves, this seemed like a pursuit of the impossible.

Weber decided to use Feynman's idea, but in simpler form by letting the two ends of a metal bar be Feynman's two masses. Passing gravitational waves would squeeze and stretch the bar. Instead of a wooden stick getting hot, piezoelectric crystals would convert the squeezing and stretching into electricity.

Joseph Weber and one of his gravitational wave detectors.

But Weber knew that even the most powerful gravitational waves must be tiny. He knew that the squeezings and stretchings in his huge metal bars would be much smaller than the vibrations induced by the thermal jiggling of the atoms they were made from.

To solve this problem Weber devised two clever techniques that are still crucial to gravitational wave detection today. One was to suspend the metal bar so that it could ring like a tuning fork or a bell. The idea was that if the thermal vibration of atoms could be made perfectly regular, like the ringing of a temple bell, you could detect tiny changes in the regular vibration. Following this idea he set about measuring the almost pure ringing tone created by the thermal vibration, and carefully searched for sudden changes in the ringing that could be caused by a passing gravitational wave.

Weber's second method was to build pairs of bars and space them some thousand kilometres apart. Then he searched for near simultaneous changes in the ringing of both bars, which could only be caused by a gravitational wave arriving at light speed, because vibrations like earthquakes travel much slower.

science news
OF THE WEEK

". . . NO REASONABLE DOUBT"

Gravitational waves detected

A 10-year effort has found radiation many thought was too weak ever to be seen

photos: U. of Maryland

Ton of aluminum proves a theory at the University of Maryland physics lab.

Einstein's theory of gravitation, called general relativity, predicts the existence of gravitational waves. These would be energy-carrying waves involving gravitational forces analogous to the energy-carrying electromagnetic waves well-known as light and radio.

Most physicists have been willing to concede the existence of gravitational waves, but they have considered the prospect of detecting them utterly remote because the waves were believed to be extremely weak.

But one physicist, Dr. Joseph Weber of the University of Maryland, believed that he could find them. An experimental program that has lasted nearly 10 years (SN: 5/27/68, p. 408) has now brought him to say: "I think there is no reasonable doubt that we see signals. I can't escape the conclusion that some of these . . . are gravitational waves."

tions are sensed by a piezoelectric crystal, and the signal put out by the crystal is processed by a sensitive and delicate electronic circuit that ultimately yields a needle trace on moving graph paper.

The needle traces a background of random noise vibrations of the cylinder. A sharp rise of the signal above the background constitutes an event that may be significant.

The idea of the experiment was to set up several detectors in different places and see if they would all record events simultaneously. An event on a single detector could be from some spurious cause, but coincidental readings on several detectors would help rule out the possibility of nongravitational causes.

After two detectors set up on the Maryland campus had experienced a number of coincident events, another was set up on the grounds of Argonne

Weber: "Can't escape the conclusion."

SCIENCE NEWS VOL. 25, 21 JUNE, 1969

In 1969 Weber caused enormous excitement amongst physicists when he announced that he was seeing signals from his detectors. For a few years he was a hero, until many physicists, who all over the world had rushed to build their own detectors, started reporting...nothing! This brings us to a sad and painful chapter in the quest for gravitational waves, but one that proved the power of scepticism in science. Before telling that story, we need to return to the astronomers.

Chapter 6

Discovering neutron stars and signs of black holes

Our story starts in the year 1054 CE. In that year, Chinese astronomers had observed a blindingly bright supernova that astronomers later identified with the Crab nebula, a bright expanding cloud of glowing gas that looks a bit like a crab. In 1965 radio astronomers discovered an unusual source of very powerful radio waves right in the centre of the Crab nebula.

Two years later Jocelyn Bell, a Cambridge PhD student, noticed a strange weak signal coming from a different spot in the sky. With persistence and skill, she revealed the signal to be extraordinarily regular radio pulses coming once every 1.337 seconds. Bell's discovery was announced in *Nature* magazine on 24 February 1968 by her supervisor Antony Hewish, Bell and other team members. They wrote: 'Unusual signals from pulsating radio sources have been recorded at the Mullard Radio Astronomy Observatory. The radiation seems to come from local objects within the galaxy, and may be associated with oscillations of white dwarf or neutron stars.'

Later that year much faster pulses, about 30 pulses per second, were discovered coming from the spot in the Crab nebula that had

Jocelyn Bell in 1967, the year she discovered pulsars.

been detected in 1965. What had they found? Alien civilisations? Astronomers dubbed them LGM for Little Green Men, but by the end of the year they had been named *pulsars*.

After much detective work, astronomers became convinced that the pulses were due to narrow radio beams coming from rotating neutron stars, a bit like a lighthouse beam scanning the sky. The radio beams seemed to be made by powerful magnetic fields, like super-powerful versions of the Earth's magnetism. Because of its coincidence with the Chinese supernova record, the Crab nebula pulsar finally proved Zwicky's farsighted conjecture from 1934: the 1054 CE supernova explosion had created a neutron star.

The discovery of pulsars was not the whole neutron star story, and it soon extended to include black holes. In 1962 astronomers

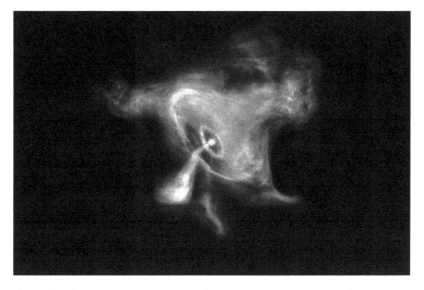

The Crab nebula as seen in X-rays. The bright spot in the middle is where the neutron star is located. It is creating a huge structure around it made of hot gas, heated by the dynamo action of the neutron star. Interestingly, the dynamo action is similar to that of dynamos used for generating the electric currents in our power grids, and the spin speed of the neutron star is similar to the spin speed of the turbines used to make our AC power.

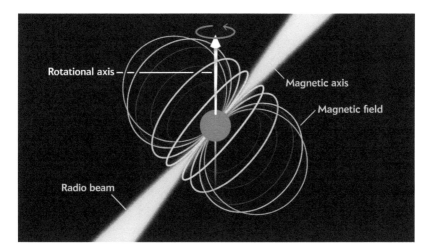

Pulsars spin around their north–south axis but, like the Earth, their magnetic poles are crooked, and create beams of electrons and light and radio waves emerging from their magnetic poles in the same way that the Earth creates auroras around the north and south magnetic poles of the Earth.

detected bright X-rays coming from a source in Scorpio, now known as Scorpio X-1, and two years later X-rays were discovered from an object now known as Cygnus X-1.

The gravity at the surface of a neutron star is unbelievably enormous. If you were to drop something from shoulder height onto a neutron star, its impact speed at your feet would be 2000 kilometres per second, thirty times faster than the fastest meteorites that strike the Earth.

Meteorites themselves must fall onto neutron stars. The neutron star's vast gravity makes them hit the surface at a few per cent of light speed. Even small meteorites would release energy much greater than that of a hydrogen bomb. The vast energy available when stuff falls onto neutron stars explained the powerful X-ray sources discovered in the space experiments.

By 1967 the Russian scientist Iosef Schlovsky had proposed that Scorpio X-1 consisted of a low-mass star overflowing its atmosphere onto a neutron star. Another X-ray source called Centaurus X-3, discovered in 1971, was thought to be a rotating neutron star being rained-on by gas from a companion star. On impact this rain of gas makes a very powerful hot spot on its surface that glows brightly in X-rays.

The X-ray emission of Cygnus X-1 was different. The energy output was very high, and observations of its blue supergiant companion star showed that it was orbiting something much more massive than a neutron star. People began to suspect that Cygnus X-1 was a giant star overflowing its atmosphere into a black hole.

It took many years to understand exactly how these energy machines work. It is a complicated process involving rapidly swirling gas that forms into an *accretion disk*, which is the disk of hot gas shown in the images opposite. The X-rays come from the super-heated gas. The gas is not neutral like the air, but a conducting plasma like the gas in a fluorescent light. As it swirls, it creates powerful magnetic fields that work like a particle accelerator and send a fraction of the infalling matter back out into space at near to light speed, while the rest of it feeds the black hole.

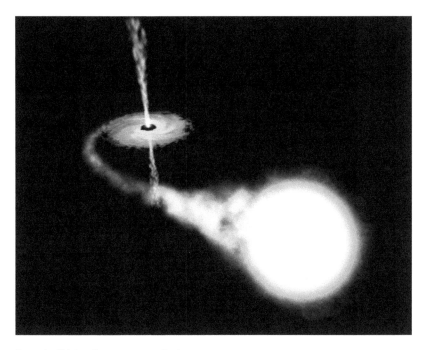

Scorpio X-1 is a low-mass star losing gas to a neutron star.

Cygnus X-1 is a giant star in orbit around a black hole about 21 times as massive as the Sun.

By the early 1970s space astronomy and radio astronomy had given us lots of evidence that the Milky Way contained plenty of neutron stars, and probably a few black holes. In 1975 Stephen Hawking and Kip Thorne had a bet about whether Cygnus X-1 was a black hole: Hawking bet that it wasn't; Thorne bet that it was. In 1990 Hawking conceded: the evidence was too great. But still all the evidence for black holes was circumstantial.

Most troubling to gravitational wave physicists was that there was still no evidence for any binary neutron stars. Because of the vast bursts of gravitational waves they could make when they collided, we badly needed them to exist!

The only other likely source of gravitational wave bursts was the formation of neutron stars in supernova explosions. But there was a problem – nobody knew how spherically the core of the star collapses. This was a crucial question because Einstein's theory of gravitational wave emission said that you only get gravitational waves from non-spherical motion. The reason is simple: the gravity of the Earth is exactly the same as it would be if all of its mass was squeezed into a smaller volume or even into a black hole at the very centre. So collapsing a star does not change the gravity outside it, hence no waves are created in a spherical collapse.

Teams of theorists around the world tried to model supernova explosions in supercomputers, but they could not make them explode! There were hopes that the newly formed neutron star would spin so fast that it would nearly break apart, but as years went by, the predicted gravitational waves from supernovae got smaller and smaller. This was rather depressing for the teams who were working on resonant bar gravitational wave detectors.

For this reason a discovery by Joseph Taylor and his student Russel Hulse in 1974 was momentous. They had been looking for pulsars using the giant Arecibo radio telescope in Puerto Rico – the one that very sadly collapsed in 2020. Because pulsars spin regularly, they are extraordinarily precise clocks. As they orbit each other, their pulses speed up as they approach us, and slow down as they recede, just like the tone of an ambulance siren changes when it passes you. This property allows any orbital motion of a pulsar to be measured extremely accurately, and the measurements can be used for precise testing of general relativity by simply measuring the time that pulses arrive.

Hulse and Taylor's pulsar orbited its unseen companion neutron star about three times every day. The neutron stars were moving in an eccentric orbit, in which they swooped in and out from about two Sun diameters to half the Sun's diameter, and were moving at between 100 and 450 kilometres per second (in comparison, the Earth orbits the Sun at just 30 kilometres per second).

Taylor carefully observed his binary pulsar over many years and shared data with theorists in general relativity like Thibault Damour, in Paris, who created very accurate mathematical models that included all the subtle things that happen when the radio pulses from one pulsar traverse the changing curved space and warped time of the other star. The pulsar beams are deflected, and the orbital patterns are distorted. Once all these effects were correctly modelled, and once they had enough data, Taylor and his collaborators could test to see if the orbit was steadily shrinking. It was! By 1982 he was able to compare his measurements with predictions and write: 'The excellent agreement provides compelling evidence for the existence of gravitational radiation, as well as a new and profound confirmation of the general theory of relativity.'

The orbit was shrinking very slowly, but in 300 million years the two stars would become as close as the binary neutron stars Freeman Dyson had imagined. Then, in their final minute, they would coalesce in the enormously powerful burst of gravitational waves that Dyson had predicted.

There was no alternative explanation to the shrinking orbit, so this system proved that Dyson's super-powerful wave bursts must happen somewhere, sometimes, in the universe. Besides proving that gravitational waves actually exist, the binary pulsar system was a laboratory for testing Einstein's theory. The results were phenomenal, and Einstein's theory passed every test. Hulse and Taylor won the Nobel prize in 1993 for their discovery.

If Einstein's third prediction, the gravitational effect on clocks, is verified Einstein's theory will dominate all higher physics...sooner or later the Einstein physics cannot fail to influence every intelligent [person].

Sir Oliver Lodge, 1920

Australia tests Einstein with radio pulsars

Parkes radio telescope, the centrepiece and symbol of Australian radio astronomy, has found about a thousand pulsars, half of all the known pulsars.

Joseph Taylor (left), who won the Nobel Prize for discovering the binary pulsar and proving that binary neutron stars emit gravitational waves, with Dick Manchester (right), an Australian pioneer of pulsar astronomy, and a leading force in a huge Australian effort to use pulsars to test Einstein's theory. Manchester was co-discoverer in 2003 of the *double pulsar*, which was even more exciting than Taylor's binary pulsar. They are photographed here in 1999.

Pulsars in Australia

Parkes radio telescope, completed in 1961, is the flagship for radio astronomy in Australia and a national scientific icon. A few days after Jocelyn Bell's discovery was published, the first pulsar was discovered by Parkes radio telescope. That was the beginning of a rich harvest of pulsar discoveries, most lead by Dick Manchester from CSIRO. The discoveries uncovered the life stories of pulsars, born spinning in the immense violence of supernova explosions, but gradually slowing down like a child's spinning top. For a hundred million years pulsars emit their radio flashes, but after that they disappear from view. Presumably they are still spinning but their lights turn off.

Parkes radio telescope made an astonishing series of pulsar discoveries, but the one that most excited gravitational wave physicists was the *double pulsar*. It was discovered in 2003 by an international team led by Marta Burgay, an Italian postgraduate student, along with Dick Manchester and others.

The exciting thing about the double pulsar was that the two stars were orbiting each other ten times every day! Such a fast orbit should give out lots of gravitational waves – enough to make the orbit shrink rapidly. Before long astronomers had measured the orbit shrinking by about 7 millimeters every day. This meant that you would only have to wait 85 million years for the two stars to coalesce in a vast burst of gravitational waves.

'So what?' you might say. Who is going to wait for 85 million years?

No one, of course! But 85 million years is a very short lifetime in astronomical terms. If you find something that is very short-lived, it tells you that they are likely to be common.

In 2003 Marta Burgay led an international team of astronomers at the Parkes radio telescope, where they detected the first known double pulsar.

Here is an analogy. Imagine you are a botanist searching a forest for a type of flower that has a bloom that lasts for just one day before it disappears. You go searching and straight away you find one. You might respond 'Wow, either they are quite common, or else I have been incredibly lucky'. You probably concluded that there were lots of them out there, even though only one was in flower today. The chance of finding something that lasts very briefly is only high if there are enough of them.

So suddenly Marta Burgay's discovery told us that binary neutron stars are likely to be much more common than we thought, because here was a very short-lived one right in our own forest, the Milky Way!

Testing the effects of gravity on time

It was all very well to be testing Einstein's theory using distant, unreachable laboratories in space consisting of radio pulsars. Apart from gravitational waves, when a pulsar beam passes near its companion, time is slowed for the pulses as they enter the gravity of the companion, and also if the beam passes near the Sun or even planets. So pulsar tests were testing the effect of gravity on time.

But still, Einstein's gravity needed to be tested here on Earth with real clocks. In 1920 the radio pioneer Sir Oliver Lodge had emphasised 'If Einstein's third prediction, the gravitational effect on clocks, is verified, Einstein's theory will dominate all higher physics… sooner or later the Einstein physics cannot fail to influence every intelligent [person]'.

NASA took up that challenge, with the idea of sending an atomic clock vertically upwards 10,000 kilometres to measure *gravitational time dilation* – the warping of time by the Earth – which could be measured by observing how time depends on altitude. NASA's rocket was equipped with a clock called a *hydrogen maser*, which emitted an extremely precise microwave timing signal. A radio beacon transmitted time signals back to the ground station. The mission, launched in June 1976 on a Scout rocket, fell back into the Atlantic Ocean almost two hours later. The timing signals clearly confirmed that time runs slower on Earth and speeds up as you leave the Earth's gravity, just as Einstein had predicted. Einstein was right yet again.

It is useful to put all this warping of time in perspective, using rough and easy numbers. The stretching of time by the Sun is about a part in a million, for the Earth it is a part per billion, for a neutron star it is one part in ten, and for a black hole it is 100%. In 1 metre of height on Earth, the stretch of time is one part in 10,000 trillion! But even this tiny effect on Earth would ruin our GPS navigators if the engineers had not made the corrections for differences in time between satellite orbits and the ground.

The double pulsar system discovered by Marta Burgay. The stars, each about the size of a city, but 40% heavier than the Sun, orbited each other ten times every day. Luckily for us both searchlight beams swept in the Earth's direction, so both could be measured separately.

The tiny effects of gravity on time on Earth are matched by its effects on space. Gravitational waves are ripples of space and they do not directly affect time. The properties of space tell us that gravitational wave signals must always be tiny. It comes about because of the enormous *stiffness* of space.

As we emphasized in Part A, space is trillions of times stiffer than diamond, the hardest known solid. The whole mass of the Sun is only enough to bend space by one part in a million, and for the Earth the bending of space is about one part in a billion – the same as their effect on time. The extreme stiffness of space means that even for the nearest, most powerful sources of gravitational waves, like colliding black holes in the Milky Way, their ripples, by the time they have travelled immense distances to the Earth, must always be tiny. However, these tiny ripples carry vast amounts of gravitational wave energy.

Of course, if you were really close to a merging pair of neutron stars where the bending of space is 10%, or black holes, where the bending of space is 100%, you might be able to feel a little judder as the rippling space passed through your body, but by the time they have crossed vast distances they will always be tiny. Roughly, the stretchings and shrinkings of space expected from the strongest gravitational wave sources might be parts in a billion billion – one trillion times smaller than the already tiny curvature of space caused by the Sun.

Thus, gravitational wave signals must always be infinitesimal ripples of space, closer to zero than any numbers we ever experience except for the exact value of zero itself – yet those nearly zero-sized ripples carry power that exceeds anything in space that astronomers have ever detected before, except for the Sun itself.

Chapter 7

Science and the human endeavour

Weber had pioneered gravitational wave detection technology, but he made the mistake of believing and not doubting. He wanted to detect gravitational waves too much! He knew that the birth of black holes could make enormous and potentially detectable bursts of gravitational waves if they came from our Milky Way galaxy.

Unfortunately, Weber lacked the scepticism necessary to find his mistakes. It took others to prove he was wrong, and when they did so, he saw it as a conspiracy to deprive him of another Nobel Prize. Astronomers pointed out that if he were right, all the stars in the universe would turn into black holes in a very short time!

Weber waged battle for many years. Time and again he showed data proving that two detectors were simultaneously being trig-gered. But when the signals went away, as they often did, he would assume that his amplifiers were misbehaving. Then he would tinker away until there were signals again. By discarding data if it did not agree with his expectations, and secretively accumulating only 'good' data, his results always looked good.

Ten groups around the world had created their own versions of Weber's detectors in a very short time. Some were significantly more sensitive, so they should have seen more signals, and the signals should show up more strongly against the noise. Stephen Hawking and his colleague Gary Gibbons proposed a more sensitive way of analyzing the data. Even with all the improvements, everyone except Weber still saw nothing.

One of my first experiences in gravitational waves was a small conference at Massachusetts Institute of Technology (MIT) in 1972 attended by two very tough physicists: Richard Garwin, who had designed the first hydrogen bomb, and Tony Tyson, who went on to play a major role in some huge astronomy projects. Garwin had built a gravitational wave detector at IBM laboratories and Tyson build one at Bell Telephone Labs. The fact that two major industrial labs had rushed into these projects showed how important gravitational wave were seen to be.

Garwin and Tyson had shared their data with Weber, and at the 1972 meeting Weber reported that he had observed coincidences between his detectors and theirs.

This seemed exciting! But then Garwin stood up and asked Weber which time zone he used for recording his time stamps. Weber replied 'Eastern Standard Time'. Garwin said 'our data is recorded in Universal time'. There was a stunned silence. The two time zones differed by five hours. So any coincidence Weber was reporting was between events in one detector happening five hours earlier than the other. The coincidences had to be spurious.

As an outsider observing science in action in the US, this was shocking. It seemed to have been a set-up, designed to publicly discredit Weber. I suspected that Garwin already knew about the time zones. Anger spilled over.

The chair of the conference was Phillip Morrison, another distinguished scientist from the Manhattan project. He had suffered childhood polio and walked with elbow crutches. In the nick of time, Morrison was on his feet. Weber and Garwin were approaching each other with fists clenched. Morrison limped forward and held up a crutch between Weber and Garwin. The tension rose but they didn't exchange blows. This is the only time I have seen scientific controversy come so close to violence.

Very slowly most physicists were convinced that there were no signals, but some signs of coincidences persisted for several years, especially between a detector in Rome and Weber's detectors. There was continuing anger and acrimony as it slowly sank in that gravitational waves had not been detected. With it came a view that, in future, the utmost care must be taken to prevent the same mistakes happening again.

Years later, in 1991, when the Laser Interferometer Gravitational Observatory (LIGO) was seeking major US government funding from Congress, there was strong opposition from Tony Tyson, who many suspected had not forgotten his unhappy experience with gravitational waves. The New York Times reported Tyson as saying, 'Most of the astrophysical community seems to feel it would be very difficult to get any important information from a gravity-wave signal, even if one should be detected'.

He went on to say, 'As a physicist I love gravity and the idea of LIGO, and the negative testimony I gave before the House sub-committee was the most difficult thing I've ever had to do...I just don't think LIGO would have much chance of achieving its goals in the next few years'.

But there was an altogether different response to Weber's announcements from another very different group of physicists. These were experimental physicists who loved challenges and saw immediately that, whether or not Weber's results were correct, there needed to be a dramatic improvement in sensitivity if we were to properly study gravitational waves. They were interested long before Weber had been discredited, and they had already embarked on a quest to build detectors a billion times more sensitive than Weber's bars. This was the beginning of a 45-year quest that ultimately detected Einstein's waves.

Western Australia reconnects

Two physicists stand out in this early stage of the development of gravitational wave detectors. The first was William Fairbank. He was a legendary low-temperature physicist who had studied the extraordinary properties of materials when you cool them down to just above absolute zero (approximately −273°C). Some metals become superconductors that always repel magnets, and electricity can flow in them forever. Helium becomes a superfluid that can flow up the walls of containers and transmit heat almost instantly.

Fairbank set up a laboratory at Stanford, where every experiment was designed to exploit the powers of these amazing materials. He had an enormous enthusiasm for science and dreamed of experiments most physicists would think impossible. Could he find out if antimatter falls upwards? Could he find out if there were a few stray quarks lurking around, left over from the big bang and not bound up into protons and neutrons? Could he measure the curved space around the Earth using a superconducting gyroscope? Could he detect magnetic monopoles – hypothetical particles consisting of a single magnetic pole? Could he use superconductors to accelerate electrons?

When he heard about Weber's experiments, Fairbank wondered if low-temperature physics could be used in detecting gravitational waves. His idea was to cool a huge metal bar to almost absolute zero, then to magnetically levitate it so it was not connected to anything that vibrates, and then use superconducting sensors to pick out gravitational wave energy pulses a billion times smaller than those claimed by Weber.

Fairbank had collected around him a team of like-minded doctoral students who were all enthusiastically working towards extraordinary goals. He was a legend in the low-temperature physics community, and he caught me up in his visions of the near impossible. He planned to build two enormous gravitational wave detectors. He introduced me to his colleague Bill Hamilton, who

William Fairbank in about 1988.

was going to build the second detector in Baton Rouge, Louisiana. The Stanford–Louisiana collaboration was going to cool the biggest bars of aluminium money could buy, coat them in niobium-titanium, the best superconducting alloy known at the time, float them on a magnetic field created using niobium-titanium superconducting coils, and measure gravitational waves. This was all proposed to be completed in the space of a few years...or so this naive Australian was led to believe.

I had believed signs in the corridors that said 'GRD for Xmas!' I thought it meant *this* Christmas, but when I tentatively asked for a Christmas break, Hamilton said, 'Oh, don't worry about those signs. They were put up for last Christmas!'

That was my initiation into the optimism of experimental gravitational wave physicists. It was a necessary quality when you were trying to do the near impossible, and inventing things that had never been done before. Fairbank, with his near-inexhaustible list of crazily difficult projects, may have infected everyone around him, or perhaps he just attracted like-minded people.

In the late 1970s and '80s Fairbank, and the larger-than-life Italian emperor of relativistic astrophysics, Remo Ruffini, were frequent visitors to Perth. Fairbank had encouraged me to return from Louisiana to Perth to work with Roy Rand, Michael Buckingham and Cyril Edwards to start gravitational wave research at UWA. Ruffini gave flamboyant public lectures and enthused everyone about black holes and neutron stars, which he had researched with John Archibald Wheeler at Princeton. Fairbank with his boyish enthusiasm would talk endlessly about his amazing experiments.

Fairbank's love of science was palpable, and he relished his visits to Australia, where one of his discoveries was the ABC's *Science Show*. If it was time for the *Science Show*, he would put his tiny transistor radio up to his ear regardless of whatever he was doing, wherever he was, and listen to Robyn Williams' interviews with scientists. One time, walking near the Swan River, a huge flock of black cockatoos were screeching and performing in the trees overhead. Our other guests were captivated by the birds, but Fairbank was captivated by the *Science Show*, a picture of concentration and reverie as the radio played loud and tinny in his ear.

Ruffini encouraged our fledgling gravitational wave team to hold a summer school to celebrate the centenary of Einstein's birth in 1979. It brought together experimental physicists interested in gravitational waves and pulsar astrophysicists, including Dick Manchester, and led our team to start a project studying pulsar evolution. Half a century after Wallal, UWA was once again deeply involved in testing Einstein.

Fairbank was the lynchpin of a strong collaboration between UWA and Stanford that included two space experiments as well as gravitational waves. Physicists John Lipa and Frank van Kann from UWA were early recruits to Fairbank's projects, especially Gravity Probe B, the space gyroscope that years later would measure the

Vladimir Braginsky. PHOTO BY USPEKHI FIZICHESKIKH NAUK.

curved space around the Earth. Later PhD graduates from UWA's gravity projects were rebuilding Stanford's bar detector when it was destroyed by the earthquake that hit California in 1989.

Sadly, this great relationship came to a close when Fairbank died suddenly on the running track in the same year. Ruffini has since held several conferences in Fairbank's honour.

While teams were building their bars, soon to be joined by another team in Rome and later one in Legnaro, Italy, another physicist, Vladimir Braginsky from Moscow, was a memorable visitor to the USA. His twinkling eyes, Russian jokes and wry humour made him the centre of attention at research-group parties and conference dinner gatherings. He had built two Weber-type detectors and was interested in the fundamental limits of precision measurement. In the basement of the vast Stalin-era wedding cake skyscraper that is Moscow State University, Braginsky had built a wonderful experiment that combined a laser with an old-fashioned giant blown-glass test tube, very fine glass fibres and a wind-up alarm-clock-powered photographic recording device. Its purpose was to test Einstein's happiest thought – the universality of freefall. This amazing apparatus yielded confirmation of the universality of freefall, to a precision of one part in a trillion, ten times better than a previous and much more elaborate experiment led by Robert Dicke at Princeton University.

On his visits to the USA and Europe, Braginsky brought a deep understanding and a shocking and ominous message, but because it was expressed in unfamiliar mathematics, it took time for the message to sink in.

Braginsky's message was actually quite simple. He pointed out that the great revelation of quantum physics – that every measurement acts back on the thing being measured – is not only valid in the domain of tiny things like atoms, electrons and photons, but it also sets limits on every precision measurement. It was a shocking discovery to realise that huge devices consisting of tonnes of metal or tens of kilograms of mirrors were actually ruled by quantum mechanics.

Braginsky's message that quantum mechanics did not only apply to the realm of atoms, but also to the macroscopic world, forced us to re-evaluate all our proposed detectors because they would be limited in sensitivity by quantum uncertainty, the intrinsic uncertainty that Einstein had so vehemently opposed.

Around this time, I had magnetically levitated a small bar of niobium, and discovered that it had exceptional ringing qualities, and hence would be less susceptible to the thermal jiggling of atoms. The problem was that if I attached any measuring device to this bar, its ringing would be damped just like a hand damps the vibration of a guitar string.

In 1975 I talked to Braginsky about this problem, and he said, 'Why don't you use a re-entrant cavity'. This was a new idea to me, although it was one that had been well developed by Soviet radar scientists working with microwaves. I got hold of the electronics from a police speed detecting radar, and used it with a home-made re-entrant cavity, quickly convincing myself that it was a good suggestion. Then I devised a way to float such a device on a magnetic field, near to a floating bar: this might be the way to build a detector in Perth.

A few years later, on a visit to Moscow, I reported on great progress in Perth, helped enormously by a very talented PhD student Tony Mann. Then I told Braginsky about our one big problem: we needed a source of very pure microwaves. He asked me if I knew about whispering galleries, where sound waves are trapped by a circular wall. This too was a new idea to me. He described a few whispering galleries around the world, including the one at St Paul's Cathedral in London, and then told me that you could make sapphire whispering galleries for microwaves. Those two suggestions of Braginsky's were the foundation for the first southern hemisphere gravitational wave detector.

Around this time Braginsky had decided to build a gravitational wave detector out of a single crystal of synthetic sapphire, because, in principle, this would ring most perfectly and suffer the least degradation from thermal vibrations. Synthetic sapphire, which is pure crystalline alumina, had been developed in the Cold War because

Two non-contacting re-entrant cavity vibration sensors. They hovered very close to a levitated niobium bar to measure its vibration. They were used in modified form on the 1.5 tonne niobium bar that was too big to levitate.

its enormous hardness made it the best material for bullet-proof windows in tanks. Yet the biggest sapphire bar Braginsky could obtain was only about 30 kilograms. A detector of this mass would be severely limited by quantum uncertainty. This motivated Braginsky to think hard about how to overcome quantum uncertainty.

1981: The heroic quest for gravitational waves

In 1919 the *New York Times* had first announced the eclipse results that made Einstein famous. In 1981 they gave a progress report that was a prescient overview of gravitational wave detection at that time. Sullivan's predictions for the future were over optimistic but the main ideas have not changed.

RESEARCHERS AROUND WORLD PLUMB THE ESSENCE OF GRAVITY

By Walter Sullivan Feb. 10, 1981

WHAT is gravity? Of the basic forces it is the least understood. No one knows for sure. But now the effort to find out — an effort that has been growing for years — has taken on heroic proportions as scientists around the world hone existing instruments and devise new ones for a variety of tests.

Physicists are virtually unanimous in their belief that, despite its seeming to be purely an attracting force, gravity has wave-properties in common with the other readily observable force, electromagnetism.

The prediction from Einstein's general theory of relativity that gravity, particularly when strong, slows time and bends space has been amply demonstrated. Watches run slightly faster on the top story of a building than at street level because, being farther from the center of the earth, they are in a weaker gravitational field. The strong gravity of the sun bends space sufficiently for stars to appear out of place when viewed along paths that skirt the sun during an eclipse.

But other predictions of the theory have not been convincingly or precisely demonstrated, such as the generation of wave motions in gravitational fields and the ability of gravity from rotating bodies to distort local space-time geometry and exert a torque on the spin axis of another body.

A passing person, car or bird presumably generates gravitational waves but they are so weak that their detection would be extremely difficult. More powerful waves, however, should be generated by events far out in space where a large amount of material is suddenly accelerated, as in the collapse of a very massive star to form a neutron star (pulsar) or black hole.

Such an event should send a ripple out through space at the speed of light, briefly — and very slightly — distorting everything in its path.

Almost all now known about the planets, stars and galaxies has been learned from electromagnetic waves — light waves, radio waves, X-rays and gamma rays. If it became possible to observe gravitational waves, knowledge of distant parts of the universe should be greatly enlarged. Gravity waves could carry information from the otherwise unobservable cores of these objects.

For 20 years physicists have sought to detect such waves, beginning with the efforts of Dr Joseph Weber at the University of Maryland. In the late 1960s, he began to report evidence of waves coming from the core of the Milky Way galaxy. The number was far greater than theorists could readily explain.

As a result, laboratories throughout the world began to build detectors similar to Dr Weber's, including some hundreds of times more sensitive. None confirmed his findings, but the belief that gravitational waves exist is still widespread, and observations now getting under way or projected are without precedent in their scope and sought-for sensitivity.

Even the most powerful waves anticipated by theorists would produce distortions whose dimensions are probably smaller than those of a single particle in the atomic nucleus. However, laser systems under development or projected should be able to measure such subtle distortions.

The detectors now in operation or in the final stages of preparation are cylinders of aluminum, niobium, silicon or sapphire crystal, designed to 'ring' in response to a passing wave. The effect is like that in which a firecracker makes a bell ring at the bell's resonant frequency although the noise of the cracker spans many frequencies.

For a convincing detection of gravitational waves, each should be observed at widely separated points to rule out local events, such as industrial tremors. Dr Weber's original observations were made in Maryland and Illinois and currently there is widespread collaboration between laboratories.

Typical of the front-line technology needed for such observations are detectors at Louisiana State University in Baton Rouge and Stanford University in Palo Alto, Calif. They are five-ton solid aluminum cylinders suspended in a vacuum and chilled almost to absolute zero. The low temperature minimizes heat-generated motions within the cylinder that limit its sensitivity.

The Louisiana cylinder is coated with a niobium-titanium alloy that, when thoroughly chilled, becomes nonresistant to electric currents. This makes it possible to support the cylinder magnetically, thus insulating it from local vibrations. According to Dr William Hamilton, leader of the Louisiana group, this is the heaviest object ever thus levitated.

Systems Use Different Materials

Detection systems using bars of aluminum, niobium, silicon or sapphire crystal are also being tested at the Universities of Western Australia, Maryland and Rochester as well as at Bell Telephone Laboratories in New Jersey.

A limit on sensitivity that had seemed insurmountable is the [quantum uncertainty principle]. A way to bypass this limitation in gravity wave measurements (known as quantum nondemolition sensing) has been devised by Dr. Vladimir B. Braginsky, who is instrumenting a 66-pound sapphire crystal at Moscow State University, and independently at California Institute of Technology by Drs. Kip Thorne and Carlton Caves.

Four institutions are seeking to observe the passing of gravity waves by means of laser beams reflected back and forth between mirrors. The beams are split in two and travel along paths typically at right angles to each other. At the ends of their round trips they are recombined.

The Interference Effect

If wave crests in one returning beam are superimposed on crests in the other, they reinforce each other. If, instead, they are out of phase, one wave cancels the other. By changing path lengths a passing gravity wave should alter this so-called interference effect.

The greater the separation of the laser mirrors the greater the sensitivity of the system. At Cal Tech, Drs. Ronald W. P. Drever and Stanley Whitcomb are building an L-shaped laser laboratory with each wing 130 feet long. Dr Drever has a similar laser system with 33-foot arms at Glasgow University in Scotland.

Laser systems are being developed, as well, at the Massachusetts Institute of Technology and the Max Planck Institute in Munich. A number of those working on such detectors believe each axis will probably have to be more than a half-mile long before gravitational waves will be detected. The MIT group, led by Dr Rainer Weiss, is designing a square laser array six miles on a side.

The installation would be insulated from earth tremors by a system that senses and actively counters tremor motions. Such an array, according to Dr Weiss, would be sensitive enough to record the gravitational effects of passing birds and aircraft, but these could be filtered out.

Observations in Space

An independent scientific panel convened by the National Science Foundation to assess efforts in this field has concluded that for maximum sensitivity — and productivity — observations with lasers or microwaves will probably have to be made in space.

Satellites could be positioned at Lagrangian points in the earth's orbit around the sun. One such point is one sixth of the way around the orbit ahead of the earth. Another is a sixth of the way behind. Objects orbiting at such points tend to be kept there by the combined effects of earth and sun gravity.

Enclosing shells would protect the satellites from drag by interplanetary gas. The shell would be kept from touching its passenger by special sensing and position-keeping systems. The satellites would be monitored to watch for wobbles in their motion.

Microwave emissions from the spacecraft Mariners 6 and 7 during their flights past Mars in 1969 were monitored for aberrations in their motion that might coincide with Dr Weber's reported events. None were detected and it is now believed the sensitivity was far from sufficient for gravity wave detection. More sensitive tests have been proposed for future space missions.

Another space test of Einstein's predictions would be made by orbiting a gyroscope under development for a number of years by Dr William M. Fairbank and his colleagues at Stanford. Relativity theory predicts very slight drift of the spin axis of a gyro in such a situation.

Experimenters believe the detectors now being built have a small chance of recording some gravitational events, particularly if a star collapses nearby. If larger detectors, such as laser arrays a half mile on a side, also fail to detect the waves, physicists are unlikely to give up.

None of the chief rivals to Einstein's theory calls for an absence of waves and there is at least one hint that they exist. A pulsar (neutron star) far out in space is orbiting another object that many astronomers believe is also an extremely dense object. Their circling is [contracting] at just the rate expected if their rotation is generating gravity waves.

If medium-sized systems fail to detect such waves, physicists can be expected to agitate for bigger ones, including — toward the end of the century — some in space.

The *New York Times*, 1981

·

Chapter 8

Cryogenic teams and laser teams

In the 20 years after Weber's spurious announcement, growing teams of physicists set about designing and building better and better detectors for gravitational astronomy. They guessed that they needed to be able to detect about a billion times less energy than Weber's detectors. The challenge was that the unimaginably small vibrations from gravitational waves had to be detected in the presence of natural vibrations one billion times bigger. They all faced enormous challenges and multiple setbacks. All were exploring the unknown, creating brand-new technologies, and discarding many that failed. Decade by decade, detectors improved by leaps and bounds, between 100-fold and 1000-fold every 10 years.

A growing community of physicists emerged and gradually condensed into two – a cryogenic team, whose priority was using very large metal bars and very low temperatures, and a laser team, whose priority was to use laser light to make measurements between widely spaced masses consisting of heavy mirrors. The first team aimed to build detectors that could fit into a big university laboratory, while the second team knew that they would have to build instruments that were much too big and much too expensive for a single university.

The reason for the difference in size was in the basic principles of their operation. The bars needed to vibrate roughly at the frequency of the expected waves, which is determined by the timescale for black holes and neutron stars to move, orbit or collapse. In the immense gravity of a star as it collapses to form a black hole or neutron star, things happen very fast. The time scale is around a thousandth of a second. When the gravitational waves they have created eventually strike a bar, it is like a sand grain bouncing off a huge temple bell. If the bounce happens in one thousandth of a second, then the best frequency for the bell is around 1000 cycles per second, although the precise frequency is not important. This means that the bars must be a few metres long, although, again, their precise length is not important.

The aluminium bar at Legnaro, Italy.

The laser detector concept was quite different. Pairs of hanging mirrors are like boats floating on the ocean. The ocean represents rippling space. You are trying to measure the ocean swell by observing the relative motion of the boats. If they were close together, they would rise up and down together, but if they were half a wavelength apart, one would be rising while the other was falling. That would be the very best arrangement. But half a wavelength for the expected gravitational waves could be a few hundred to 1,000 kilometres. Clearly further apart is better, but there is a practical economic limit set by the cost of building long straight vacuum pipes for sending laser beams without disturbances from the air. The compromise length chosen was 3–4 kilometres.

The cryogenic team, which began with Stanford University and Louisiana State University, quickly grew to include the University of Rome and the University of Western Australia, and then grew further in Italy to include detectors at Frascatti, Legnaro and CERN. The 1989 earthquake that caused Stanford to suddenly drop out was an interesting lesson in inertia. When it struck, the enormous and delicately suspended aluminium bar remained stationary. But the large structure of pipes, tanks and sensors, well attached to the ground, suddenly moved about 1 metre horizontally. The apparatus was destroyed as if hit by a giant battering ram. The battering ram was the bar, and which part was actually moving is just relativity!

Cryogenic cooling had many advantages but also many difficulties. The advantages of cooling, which reduces the jiggling of atoms, not only reduces thermal vibration, but also causes the acoustic ringing properties of many materials to become much better, all of which makes it easier to distinguish between gravitational wave signals, which come suddenly, and thermal vibration, which changes slowly. The third advantage is that sensors based on superconductors can be vastly more sensitive than regular electronic sensors such as microphones. The disadvantage was that each bar had to be surrounded by a massively complex cooling system, and like most refrigerators, they made unwanted vibrations.

The laser group began to develop systems where laser beams acted like Feynman's stick joining widely separated masses that were themselves mirrors. The mirrors follow the stretching and shrinking of space as gravitational waves travel past, and the laser measures their motion. The easiest way to do this is to use pairs of mirrors in an L-shaped configuration, because nature gives us an ideal method, called interference, for detecting such changes. Interference gives the coloured pattern we see in soap films, when photons of certain colours are either reflected or pass through the soap film according to the film's thickness. Tiny changes in thickness make dramatic changes in colour. The same process can be duplicated using laser light, so that its intensity depends on the relative distance the light travels to and from distant mirrors at the ends of a giant L-shaped vacuum pipe.

The stretching and shrinking pattern of a gravitational wave makes you alternately get shorter and fatter, then taller and thinner,

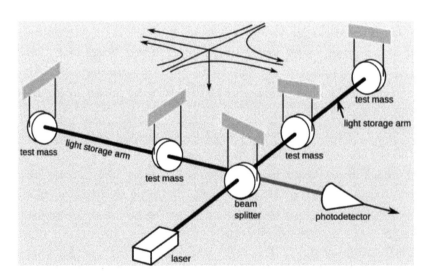

The diagram shows the main mirrors of a gravitational wave detector hanging freely on wires. The pattern of the gravitational wave is shown coming in from above, causing the stretching of space in one direction and shrinking space in the other direction. A moment later as he wave passes, the pattern reverses.

perhaps hundreds of times each second. The same happens to the L-shape. The brightness of the light at the apex of the L-shape tells us the relative length of the arms. The advantage of this technology is that the sensitivity to distance changes doesn't change as we make the device longer. But the total stretching of space does increase with distance, so a longer instrument is always more sensitive. The disadvantage is that the devices need to be huge and are extremely complicated.

Chapter 9

Exploring the science of measurement

Supersensitive PA systems

All the teams building gravitational wave detectors were aiming to measure motion as small compared with an atom as an atom compares with a person. In numbers, a person is very roughly a metre in size, whereas an atom is about 10^{-10} metres; that means that it needs a line of 10 billion atoms to stretch from your head to your feet. The gravitational wave detectors needed to measure about 10^{-20} metres – that is, one ten billionth of an atom. Ten billion of these units stretch from one side of an atom to the other!

How could such a measurement be possible when everything is made of atoms and the atoms themselves jiggle around with their own heat? The answer is by averaging over a surface. If you measured one spot, it indeed would be jiggling wildly, but average over a trillion atoms and the result can be very stable.

Besides the thermal vibrations, our whole planet is vibrating constantly. The ground under our feet is vibrating with an amplitude about 10,000 times larger than an atom. If we could see the ground like we can see the surface of an ocean or a lake, we would see a constant pattern of waves of tiny amplitude but long wavelengths racing across the ground at between 300 metres per second and 3 kilometres per second. The height of these seismic waves is about a million times smaller than the waves of the ocean, but their speed is hundreds of times higher. They are quite imperceptible to our senses, but they are easily detected with seismometers.

Huge, complex vibration isolation systems [see image] can suppress these seismic waves. Once this is done, you have to consider how to measure the motion of the masses as they respond to the gravitational waves, without disturbing the masses.

An example of a vibration isolator, this one developed at the University of Western Australia for use at the Gingin high optical power facility.

If you have ever experienced a bad PA system, you will understand some of these problems. The first problem you might have noticed in PA systems is a loud and unpleasant hiss of white noise coming from the loudspeakers – it can be loud enough to drown out soft voices or music. You might have heard awful noises from the wind blowing over the microphone. Even worse, if a person has a soft voice and someone turns up the volume, the whole PA system can start to whistle or scream. This is called feedback. It happens when the sound from the loudspeaker reaches the microphone, gets amplified, and then returns to the speaker even louder. Sometimes it can build up into an awful crescendo. When using the internet you might have experienced delayed feedback, which is like a loud echo that makes it completely impossible for you to speak at all.

All of these common phenomena also afflict gravitational wave detectors. You can think of gravitational wave detectors as PA systems for listening to the ripples of space. They are very much like PA systems, only vastly more sensitive, and the voices are very very soft.

The problems of measurement seem to be universal – however you measure and whatever technology you use, the same sorts of problems are encountered. We encountered them when we built Niobé, the niobium bar gravitational wave detector (named after the Greek god, the daughter of Tantalus), and again when we developed laser technology.

At UWA we first built an extremely sensitive microwave powered 'microphone' for our niobium bar (see image on page 151). It consisted of a tiny resonant superconducting structure that was tuned by proximity to the surface of the niobium bar. The vibration of the bar changes its tuning like a hand can modulate the tone of a whistle. To be sensitive, our microphone needed a stable source of microwave photons (photons similar to those used for 5G mobile phones). The microwave source needed to be extremely stable – we needed it to make a perfect and pure note, constant in both frequency (i.e, with constant pitch) and with intensity (with constant loudness), because a fluctuating frequency would make it hiss and changing loudness would disturb the oscillations of the bar itself.

No suitable source of microwaves existed so we had to invent something. The device we built was called the sapphire clock. It used the remarkable properties of whispering galleries that Vladimir Braginsky had introduced me to. I had come back from Moscow with Braginsky's suggestion scribbled on a serviette, and immediately contacted a little US company that specialised in fabricating sapphire. Within a few years we had improved on the concept, patented the new idea and had sources of microwaves that allowed us to achieve the sensitivity we needed, as well as demonstrating their capabilities for both time keeping and supersensitive radars. We developed clocks for use in radio astronomy, but their greatest value was in giving our gravitational wave detector the lowest level of noise ever measured.

I can't resist adding one last side story: the third postgrad student to work on the sapphire clock moved to another university, and suddenly our team started noticing public claims that he had invented the sapphire clock. A tiny Google search reveals that alternative facts are alive and well in some sectors of the science community. It was invented and patented at UWA in the 1980s.

The building of Niobé at UWA allowed us to uncover a very important phenomenon, not of great importance for our own gravitational wave detector, but one that would be very important for the huge laser gravitational wave detectors that would supersede our metal bars. The problem was a new and very subtle form of feedback called *parametric instability*, which is also a beautiful illustration of quantum physics happening on a very large scale.

Our microphone for reading out the vibrations of the niobium bar had been designed to 'suck' sound energy out of the niobium bar. What we discovered was its ability to simultaneously inject sound into the bar at a harmonic of the bar frequency, so that a certain pitch would very slowly build up into a terrible loud ringing.

We explained our discovery in terms of the energy quanta that Einstein had discovered in 1905 and 1907. We only told a part of this history in Part 1 of this book, which was Einstein's proof that light and all electromagnetic wave energy comes as photons. In 1907

Einstein showed that sound and heat must also come in tiny packets. The modern name for these packets of sound is *phonons.*

In our microwave-powered microphone, each microwave photon from our sapphire clock was being split into two energy packets. One was a new slightly lower energy photon, and the other was a sound energy packet – a phonon. The sum of the frequencies of the new photon and the phonon was exactly equal to the frequency of the original photon. The ringing of the bar was a result of all those unwanted but unavoidably created phonons building up inside the bar.

It was not a huge problem for our project because, fortunately, we had a brilliant Russian physicist in our team called Eugene Ivanov. He was a master of control systems and was very quick to understand the problem. Then he created a simple solution by adding a feedback system to suppress the instability.

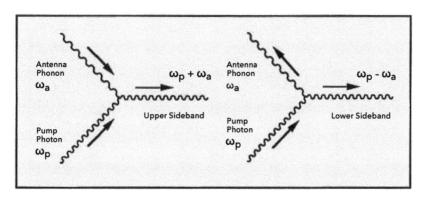

This diagram represents how photons and phonons interact when microwaves are used to measure vibrations of a bar gravitational wave detector. No picture of a photon or a phonon is ever correct because how they act depends on how you measure them. As represented here, they act a bit like billiard balls bouncing off each other, but they also have a frequency represented by the squiggly lines in this diagram. One thing we can say for certain is that the frequencies add and subtract exactly. Reading the vibrations of a bar combines phonons with photons to create new energy quanta shown by the arrows going away from the vertex. New photons are always created – they are called sideband photons, but phonons either enter the bar or leave the bar. If too many enter the bar, they can create uncontrolled ringing called parametric instability.

While we were investigating this new parametric instability phenomenon in 1993, Caltech and MIT were designing their huge laser-based detectors called Laser Interferometer Gravitational Observatory, or LIGO. No one imagined that parametric instability would be a problem for these detectors.

It was several years before Vladimir Braginsky realised that the parametric instability we had observed in Niobé could happen in laser-based detectors. In this case, a laser photon would split into a lower energy photon plus a phonon that was a quantum energy packet of mirror vibration. As you increased the intensity of the laser light, more and more phonons would enter the mirrors and could build up uncontrolled vibration. This could occur at many possible frequencies.

At UWA we immediately understood that parametric instability would be far more serious in laser detectors, because the huge mirrors used to reflect the laser light had literally thousands of harmonics of sound vibration that could begin ringing, powered by laser photon energy. Chunnong Zhao and Ju Li (pronounced joo-lee), with Polish PhD student Slawek Gras, did a very careful analysis and made a worrying prediction: instability would be inevitable in the second stage of the LIGO project called Advanced LIGO. The Advanced LIGO was the first detector designed to be sensitive enough to detect predicted gravitational waves from pairs of coalescing neutron stars. Our predictions identified a serious threat to the project.

The synergy between the technology of our niobium bar and the laser detectors had given us the insight. We felt confident in our predictions. All the conditions needed to increase the sensitivity of the billion-dollar laser detectors were also the conditions we had on the niobium bar where we had discovered the instability problem. We had put our reputation on the line. Now we needed to prove that we were right!

This was the beginning of a ten-year struggle, first to prove parametric instability, then to convince our LIGO colleagues that it was a serious threat, and finally to make a big contribution to the detection of gravitational waves through our accumulated experience.

Before we follow the lasers any further, we need to look at the outcomes of the bar detectors.

Niobé comes online

Niobé: Monitoring the Milky Way

The Australian Physicist, 10 June 1994

The Perth antenna is quite different from any other antennas in the world. It is the only niobium bar, the only successfully operated antenna with a parametric transducer and the only antenna with a non-contacting readout. In recent years referees, while applauding our perseverance have said we should abandon our cantilever spring catherine wheel, abandon the parametric transducer and abandon the non-contacting coupling. However, without these features, we might as well have abandoned the niobium bar itself and gone over to a lower quality aluminium bar. Now our perseverance has been vindicated. We have achieved the lowest noise temperature ever recorded in a gravity wave antenna. We are continuously monitoring the Milky Way, and hoping for a gravitational collapse!

Late in this decade large scale laser interferometer gravitational wave observatories will probably supersede resonant bars. With broad bandwidth and greater sensitivity they should make gravitational wave astronomy a reality. In the meantime the bars have a modest chance of making the first detection of a gravity wave.

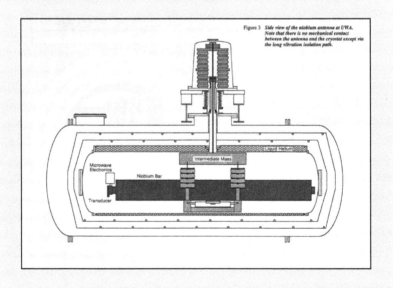

Figure 3 *Side view of the niobium antenna at UWA. Note that there is no mechanical contact between the antenna and the cryostat except via the long vibration isolation path.*

Editorial, *The Australian Physicist*, 1994
By Jack Kelly

Gravity and personality

A piece of university built equipment which takes seventeen years to become operational and, even when it is working, may not have anything turn up to be measured for further years, sounds like the sort of thing that sends PhD supervisors prematurely grey and drives experimentalists to drink. Fortunately Western Australia provides some splendid wines for just such an emergency. This drawn out saga deserves a happy ending and you will see that such is the case by reading the article on the Perth gravity wave antenna on page 134 of this issue. It is written by David Blair and features a cast of thousands. They have overcome enormous difficulties to produce the lowest noise temperature ever recorded in such a device and on the way made significant contributions to a number of other areas mentioned in the article.

Those of us outside the field have, over the years, heard rumours, and occasional facts, of what was going on in Perth and admired their persistence. Through the eighties, when so many people were jumping up and down shouting: 'What do we want?" "Everything!" "When do we want it'?" "Now!", it was nice to know that a really worthwhile project with a much longer time scale than was fashionable was still moving forward.

There is obviously a good book in this. I hope someone is recording oral history from the many involved and preserving the details. One must particularly admire those capable of persuading a University Investment Committee to lend the money for 1.5 tonnes of pure niobium, a metal they had probably never heard of. This achievement is all the more noteworthy at a time when several Perth entrepreneurs were offering what looked like much more rewarding investments. Surely a classic example of investment in physics paying off, or at the very least avoiding a serious financial loss. Let us hope that the University remains prosperous enough not to need its niobium back, at least not before significant gravitational waves have passed through Perth.

The International Gravitational Events Collaboration

The five cryogenic bar detectors reached high sensitivity in the 1990s. They worked together as the International Gravitational Events Collaboration. Four of them used aluminium bars and superconducting quantum interference device sensors that were bolted onto the bars. Only Niobé at UWA used a non-contacting sensor. All had similar sensitivity to gravitational waves. Niobé was a bit smaller than the others but this was compensated for by having the lowest noise energy ever measured at that time.

The bars collected data and had precision timing systems so that all data were accurately time-stamped. They shared their data so that everyone could independently search for multi-detector coincidences that might be due to gravitational wave events in the Milky Way. In all there were five detectors, 25 years of construction for each one, and a collective undertaking by more than 100 doctoral students and a similar number of scientists and technicians.

With vast amounts of data generated, it needed equally vast amounts of effort to analyse results from instruments that did not always perform well. Teams across the world had had to develop new methods of analysing data, studying the statistics of noise and methods for comparing signals from detectors that were each individual and had certain peculiarities to their noise. To avoid unconscious and conscious bias, we had added random time offsets to our time stamps, so that no one knew if any coincidences were real until the true time was revealed. Most of the analysis techniques were later applied to the next generation of laser interferometer detectors.

At UWA the hero of the data analysis was Ik Siong Heng, a Singapore-born Perth physicist who dedicated years to analysing data from Niobé, then analysing it with data from individual detectors, and finally contributing it to a global data analysis challenge.

Occasional mystery signals kept life interesting, even exciting, but the rigors of statistics and lots of healthy collective scepticism ensured we did not repeat Weber's mistakes.

Every gravitational wave detection team was dependent on excellent data analysts. Many young scientists were deeply involved. One of the leaders was Pia Astone from the University of Rome La Sapienza. In 2001 she presented a review of all our detector results at the Fourth Eduardo Amaldi Conference that was held in Perth in 2001. This conference was held in honour of Amaldi, who was both a founder of CERN and the founder of Italian gravitational wave research effort.

At the Amaldi conference, Astone reported that no signals had been found, but the collaboration had set very strong upper limits on the number of gravitational wave events and their strength in the Milky Way.

The setting of upper limits is of huge importance when searching the unknown. Each upper limit is a result in itself. Today, when gravitational wave detectors are in the news for discovering many definite sources, the less newsworthy science is about the many other predicted types of sources that have not yet been detected. Each result is an upper limit that specifies the maximum strength and how often events could be occurring without having been detected.

The International Gravitational Events Collaboration proved that black holes are not being born very often in the Milky Way. This was not really a surprise, but some of us had surely hoped that most of the mysterious missing mass of our galaxy – the dark and unseen matter – might be a population of black holes.

Resonant
mass
detectors

Pia Astone

4th Amaldi conference

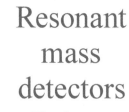

Perth July 8-13, 2001

Our Milky Way galaxy is very much alone in our part of the universe. The Andromeda galaxy, our closest large neighbour, is at a distance of about 3 million light years. Within 20 times that distance of Andromeda, we reach the Virgo cluster of 1500 galaxies, where black hole births could reasonably happen ten times per year. The teams that built the bars knew that signals from these galaxies would be quite out of reach. The only option was to build detectors 1000 times larger, and that meant laser interferometers.

Some of the bars continued operation for ten more years in the hope of a galactic supernova, while others like Niobé found a new home on display at the Gravity Discovery Centre.

Was this all a wasted effort? I think not. We had learnt the physics of ultrasensitive measurement. We had had to verify that any small excitation observed in our cold and quiet bars – some joked that we had created the quietest and coldest places in the universe – were not caused by cosmic rays or earthquakes. Niobé had ridden

Niobé in the Gravity Discovery Centre.

out one quite significant earthquake that made it swing around for hours, but that did not affect its exquisite performance. Our measurement limits corresponded to tiny amounts of energy – the smallest amounts of energy ever measured. We had learnt how measurement acts back on the system being measured and even how quantum uncertainty can be suppressed. We trained a huge team of gravitational wave scientists, many of whom brought their knowledge and skills to the new challenge of making laser interferometer detectors work as planned.

Measuring with photons

The idea of using photons to measure gravitational waves allows powerful insights into the measurement process. One of Einstein's most important revelations was that photons have momentum. For school students we describe this quality as *bulletiness* because, like bullets, photons can push things or break things. But photons also have *waviness*: this is the quality of two beams being able to interfere, like light in a soap film, where photons appear or disappear according to the length of their paths.

Laser interferometer gravitational wave detectors harness both the bulletiness and the waviness of light. The waviness makes the number of photons per second at the output depend on the length of the detector arms. At the output, the photons get detected by a sensor in which each photon uses its bulletiness to kick an electron. All the kicked electrons create an electric current proportional to the light intensity.

But photon arrival is governed by probability: strictly every measurement follows the rules of chance, and like any chance situation, you only get a good estimate if you have many samples. For light, this means you need a lot of photons – i.e. high light intensity.

The statistical fluctuations in light intensity create a background noise that is like the hiss in the speakers of a PA system: it is a direct result of the randomness of light. Physicists call the noise from this randomness *shot noise*. It sounds a bit like lead shot or tiny pebbles being poured onto a metal sheet.

The logical deduction from this is that all laser interferometers should use the highest possible laser intensity to minimise the effect of photons arriving randomly. But this is not the case. The intensity does need to be very high, but we must not forget the bulletiness of the photons. The greater the light intensity, the more photons are hitting the mirrors. If every photon is like a bullet, then when the photon reflects, the mirror recoils. But the photons are following the rules of chance – sometimes there will be more photons, other times less, but the absolute fluctuation in number always gets bigger the more photons you have. Mathematically, the fluctuations increase as the square root of the number of photons. For 100 photons, the fluctuation size is about 10 (because 10 × 10 equals 100), but for one million photons it is about 1000. Although 1000 is only 0.1% of one million, the fluctuating push is 100 times more than with 10 photons.

Physicists have two names for this recoil effect: radiation pressure noise or back action noise. The name *radiation pressure noise* refers to the pressure that light exerts on everything. *Back action noise* refers to the problem that the measurement system always disturbs the measurements.

I wonder how Einstein would have felt about this? Our detectors are ruled by the laws of chance. The laws of chance tell us that if you have too many photons, the detector's mirrors will swing around like crazy, driven by the photons hitting them, but if you have too few photons there will be lots of background hiss.

God surely does play dice! The laws of chance rule the universe. They set limits to how well we can measure, but with ingenuity, this can be well enough. The waves that Einstein first predicted, and then doubted, are created by the black holes that Einstein was sure could not exist in physical reality. Yet they were finally measured. They proved the extraordinary predictive power of humankind's greatest intellectual achievement: Einstein's theory of general relativity.

Laser interferometer concepts take shape

While the bar detectors were being constructed, the parallel laser projects were still developing. By the 1990s the bars were being readied for operation. By this time the laser projects had made some crucial discoveries that would allow them to plan the transition from laboratory-scale science to big science.

The small-scale prototypes of laser interferometer gravitational wave detectors built during the 1980s aimed to test all aspects of real detector technology. Their sizes had been set by building sizes and sites available – 10 metres at the University of Glasgow, and 40 metres at California Institute of Technology and others in Germany and Japan. They were all much too small to be real detectors, but while their experiments were progressing, design teams were planning a 600-metre-long detector in Germany called GEO, a 3-kilometre-long detector near Pisa called Virgo, while Caltech and MIT scientists proposed building two 4-kilometre-long detectors they called LIGO: the Laser Interferometer Gravitational Observatory.

The prototype detectors had proved that it was possible to hang mirrors on delicate suspensions with ringing properties sufficiently good that the noise from thermal vibrations was small enough. They had developed magnetic methods to gently adjust the position of mirrors into precise orientations. They had shown that a laser light could be made extremely pure and stable, and that beams could be split perfectly into two parts by a beam splitter and be directed to the distant mirrors and then recombined so that interference would allow comparison of the length of each arm.

The lasers and mirrors needed to work together in perfect sync. This meant locating and controlling the mirror positions to much less than a nanometre – the size of an atom. In detector parlance, the mirrors were *locked* to the laser beams. The locking was achieved using servo control systems that would correct for disturbances, rather like a self-driving car.

All of these techniques had been proved, but with them had come a couple of enormous breakthroughs. These concerned the laser power needed for detecting the expected tiny signals. The big detectors would never be sensitive enough unless their laser power was enormously high. This was needed to reduce the statistical uncertainty of the randomly arriving photons.

The problem of needing very high laser power was overwhelming. Stable lasers of the type they needed had a power of a few watts, but calculations said that almost a million times more photon power was needed. There were rumours about the military designing megawatt lasers for shooting down missiles, but such lasers could have no place in a precision measurement laboratory.

The story of these crucial discoveries is a fascinating story of various personalities.

The laboratory at Glasgow University was led by Ron Drever. He was a memorable character who reminded me of Dr Doolittle. He was short, slightly plump and enormously enthusiastic. He had a broad Scottish accent and the fastest speaking rate I had ever encountered. Drever had tried to duplicate Weber's results. He built a novel but equivalent detector that used two short aluminium bars. He sandwiched piezo sensors between them. (Piezos are commonly used in microphones and gas lighters; they directly turn vibration into electricity.)

Ron Drever. PHOTO: CANDID CAMERA RINALDI,
COURTESY AIP EMILIO SEGRÈ VISUAL ARCHIVES.

While the bar groups were asking how to make Weber-type bars much more sensitive, Drever, from the start, was thinking about separate masses, closer to Feynman's original concept of masses linked by a stick. He began to ask if it would not be better to use lasers to measure vibrations instead of the piezoelectric sensors. Drever was not the first to think along these lines. In 1962 Michael Gertsenshtein and Vladislav Pustovoit in Moscow suggested using a laser interferometer to 'monitor the relative motion of freely hanging mirrors in response to a gravitational wave'. Over the next few years two very different physicists would make huge contributions.

Robert Forward had done a doctoral thesis building one of Weber's bars. He went on to join Hughes Aircraft Laboratories, where he seemed to be able to dabble in whatever exotic physics took his fancy. He was a colourful and wildly creative character. At Hughes, Forward constructed the first laser interferometer gravitational wave

Robert Forward.

detector and recognised the critical limitation set by the statistics of photons. He also invented a very clever concept for measuring hidden masses by their gravity, which was the precursor to the Falcon Gravity Gradiometer now widely used for airborne mineral exploration in Australia. UWA is deeply involved in improving this amazing instrument.

Some other proposals by Forward were much more speculative, such as a proposed antimatter drive for rockets and the concept for a 20% of light speed spacecraft called *Starwisp*, which re-emerged 30 years later under the banner Project Starshot, led by Stephen Hawking. Forward was fascinated by neutron stars and was the first to calculate the gravitational waveform from two neutron stars spiralling together that Freeman Dyson had anticipated a few years earlier. Forward described the first of his 11 science fiction novels, *Dragon's Egg*, as 'a textbook on neutron star physics disguised as a novel'.

Around this time, Rainer Weiss at MIT was studying the theoretical limits of laser interferometer gravitational wave detectors. He submitted a proposal for funding a research program that was sent for peer review to Professor Hans Billing, one of the disappointed physicists who had built a Weber-type bar in Munich. Billing was excited by Weiss's proposal and set his own group working along similar lines. Weiss, for some reason, did not get funding, but the Munich group went forward and collaborated with the Glasgow group. The collaboration grew into a British–German project for a mid-scale detector called GEO, which pioneered the most advanced detection technologies, while Weiss went on to share the Nobel Prize for the discovery of gravitational waves more than forty years later.

Drever, meanwhile, began experimenting with configurations of laser beams that could increase their sensitivity. The problem that Drever well understood was that the wavelength of laser light is about a millionth of a metre, but gravitational wave detectors needed to measure a trillionth of the laser's wavelength. Light intensity changes from interference can measure a thousandth of a wavelength, but it was far-fetched to think that you could measure one trillionth of a wavelength.

The early approaches to building up sensitivity were geometrical: find ways that light could zigzag many times back and forth between the mirrors. For example, if the laser beam zigzagged 100 times between mirrors, then if the mirrors changed their spacing by a certain amount, the laser beam path would be stretched by 100-times that amount. There turned out to be a fatal flaw in this approach, proved by the German group. It was that light scattering from the not-quite-perfect mirrors could get into the final bounce beam, smearing out the interference: the problem got worse the more bounces you had. There had to be another solution.

In popular culture, Scottish people are often known for their thriftiness: 'dinnae waste a morsel'. Ron Drever was a brilliant experimentalist, a great improviser and a true Scot. An example of his improvising was the rubber toy cars from toy shops that he

Jim Hough (left) and Ron Drever (right) in front of their split aluminium bar.

re-purposed as springy pads for absorbing vibrations. But his huge contribution to the field was when he applied Scottish thriftiness to wasted photons.

Drever's laboratory at Glasgow first pioneered the idea of putting pairs of very carefully positioned mirrors in the arms of a laser interferometer. One mirror must be partially transmitting so that light can enter, and then the light can resonate back and forth between the mirrors. Two things happen when you do this. First, the light intensity builds up between the mirrors in the same way that a person can build up the swinging of a pendulum with many small pushes. The light can become hundreds or thousands of times brighter than the incident light. Second, when two arms like this are put together in an interferometer, its sensitivity to the position of the mirrors is enhanced by the same factor.

The concept of resonating the light between mirrors was not new in optics, but the idea that it could actually be made practical for hanging mirrors in gravitational wave detectors was revolutionary. But this was just the first piece of photon thriftiness to come out of Glasgow.

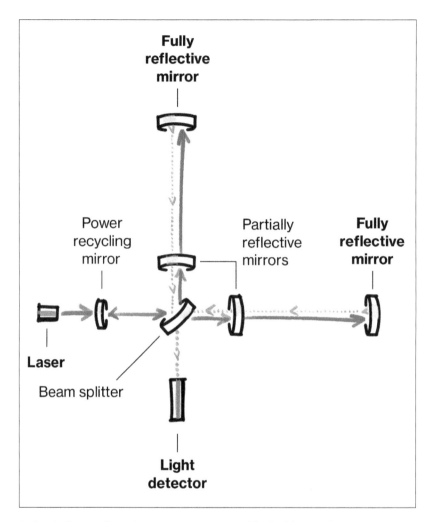

A sketch showing how the mirrors are arranged for building up laser power in a gravitational wave detector. In addition to the mirrors at the end of the long arms, two mirrors are placed in each arm, and another mirror is placed in front of the laser. The addition of these three highly reflecting mirrors makes the total light intensity increase hundreds or thousands of times.

Drever's second revolutionary idea happened in 1983 when he and I were lecturers at summer school on gravitational waves at Les Houches in the French Alps. I was talking about vibration sensors for bars and Drever was talking about laser interferometers. In his rapid-fire lectures, Drever pointed out that the best sensitivity of a laser interferometer was if it was tuned so that no laser light came out of the interferometer except for the dim flicker of light when a gravitational wave disturbed it. He lamented that almost all the photons went back towards the laser, where they were wasted. Because they can disrupt laser operation, they are usually dumped by a device called an optical isolator.

Over drinks that evening he began discussing ways it might be possible to recycle those wasted photons. He was animatedly scribbling little sketches on napkins. In the morning he showed me some very roughly scribbled arrangements of mirrors and laser beams that showed how photons could be re-injected and re-used. This was the birth of an idea called *laser power recycling*, which allows the laser light to be reused many times over, thereby building up the total laser power in the interferometer by factors that can be hundreds or thousands.

Several years later, Drever's collaborator Brian Meers showed that you could also recycle some of the photons in the output of the interferometer. This could actually enhance the signal strength by resonantly building up the photons.

Drever's ideas revolutionised the concept of the laser interfero-meter from one that needed impossibly powerful lasers to one where the laser power needed was built up inside the detector from a relatively modest laser that was quite practical. The magic of laser optics meant that all you needed to convert a simple and rather insensitive interferometer into a highly sensitive one was the addition of four partially reflecting mirrors: one in the input, two in the long arms and one at the output.

But to turn these ideas into reality was a massively complex undertaking.

Firing up laser gravitational wave detection in Australia

Instigated by Remo Ruffini, in 1988 our team at UWA played host to the Fifth Marcel Grossmann Meeting on Relativistic Astrophysics. The conference series was started by Ruffini and named in honour of Einstein's friend, who had helped him with the mathematics of curved space. It brought together physicists interested in all aspects of gravitational waves and astrophysicists interested in black holes and neutron stars. Joseph Weber attended, still proclaiming that he had discovered gravitational waves. Also present was James Hough, the new leader of the Glasgow group, and Ron Drever who had moved to Caltech to spearhead a new experimental program in the USA. Hough suggested we should think about a possible UK–Australia collaboration to build a huge laser interferometer but warned that this might be impossible because of the UK joining the European Community. He was right, but his suggestion lit a fire in Australia, and without it Australia might not have been part of the eventual discovery of gravitational waves.

What we saw at the 1988 Marcel Grossmann meeting was concepts for huge gravitational wave detectors being discussed for Europe, Japan and the USA. Momentum was building, and a new team at Australian National University began discussing with the UWA team the idea of joining forces to build a laser interferometer gravitational wave detector in Australia. At this stage UWA was still preparing to operate the only gravitational wave detector in the southern hemisphere. Long-term operation with four other detectors in the northern hemisphere was still in the future (from 1993 to 2001).

1995: Australian Consortium for Interferometric Gravitational Astronomy

Discussions in Australia progressed, and in 1995 physicists at Australian National University, University of Western Australia, University of Adelaide and CSIRO in Sydney set up a national collaboration called ACIGA, the Australian Consortium for Interferometric Gravitational Astronomy. The idea was to develop laser interferometer technology for a future Australian detector, and to collaborate with international projects.

Everyone recognised the need for an Australian laser interferometer detector. The Earth is about 40 milliseconds in diameter in light speed or gravitational wave speed units. This means that there must always be a delay of about 40 milliseconds between a wave striking one side of the Earth and the other. If the arrival of signals is timed accurately enough, the timing at each detector allows the direction of the incoming wave to be pinpointed. Australia is

The first power recycling interferometer in Australia, now the entranceway to the Gravity Discovery Centre, Western Australia.

optimally located compared with detectors in Europe and the USA to have a maximum time delay, and hence makes the best possible contribution to pinpointing sources of gravitational waves.

The Australian team began collaborating with the teams behind three huge detectors planned for the USA and Europe: the two LIGO detectors in the USA, and the Virgo detector in Italy that was developed as a collaboration between France and Italy. Australia was ready in 1997 when the formation of the LIGO Scientific Collaboration allowed scientists from around the world to participate in LIGO. Many Australian scientists, particularly from UWA, were invited to spend lengthy periods in Italy helping to design the Virgo detector, while many more worked with LIGO, especially on the planning of the next version of LIGO: Advanced LIGO.

At UWA we built an 8-metre interferometer using a big vacuum system donated by Woodside Australia. Here Ju Li created our first vibration isolators and Chunnong Zhao created the first power recycling interferometer in Australia. Today, that interferometer is the entrance archway for the Gravity Discovery Centre.

Australian Centre for Precision Optics

Australia had one major asset: CSIRO's Australian Centre for Precision Optics, ACPO. A very strong optics group had grown up over the years in the National Measurement Laboratories in Lindfield. They were absolute masters in polishing mirrors and lenses for a vast range of applications where state-of-the-art optics was needed. Their technology was used in space science, measurement science, microscopy and telescopes. If anyone needed optics, ACPO could make it if it was humanly possible, and were the best in the world at what they did. Their customers were scientists all over the world.

We talked about their expertise to our colleagues around the world. The outcome was that LIGO hired ACPO to fabricate their test mass mirrors. These were to weigh 30 kilograms each and be the most perfect mirrors ever created, with their spherical curved mirror profiles precise to near atomic dimensions. It was an enormous credit to Australia that our optics was at the heart of LIGO.

Soon afterwards, ACPO was asked to create even better optics for Advanced LIGO. Then a bombshell hit.

One of the key components of LIGO were the 30 kg test mass mirrors. These were polished in Australia by CSIRO.

A LIGO mirror – no longer made in Australia.

In 2014 the Abbott government slashed CSIRO's budget, and the entire ACPO centre was closed down. Half a century of accumulated knowledge and skill was trashed. One of the crown jewels of Australian science was destroyed by a narrow-minded, ill-considered political decision. This was a loss to Australia and a loss to the entire world.

In 2016 when the detection of gravitational waves was announced, CSIRO claimed credit for their contribution, while failing to mention the fact that they had disbanded the labs and the workforce that had made it all possible. Shame!

1999: Gingin site for gravitational astronomy

The Western Australian government had provided funds for a feasibility study for a gravitational wave observatory. We had found an ideal site in the Shire of Gingin at a place called Wallingup Plain. The first thing we did was make contact with the traditional owners of the site, the Yued people and the Nyungar people. We held a succession of bush meetings, sitting under trees in the flat pristine sand plain that was only accessible by four-wheel drive vehicles. John Sandeman, the senior leader of the Australian Consortium from ANU, joined our final meeting, where we drew up an agreement in which we promised to preserve the environment, preserve the sacred *nyutsia floribunda* Christmas trees, and involve Aboriginal people in activities on the site.

In 1999 we asked the Western Australian Government for seed funds to build the labs we would need for a research centre and a prototype interferometer. The response was 'you should be able to do more for Western Australia than just a few boffins beavering away in the bush!' I had an instant answer. 'We could build a public education centre.' There were nods of approval while a voice in my mind said 'This is crazy!'

Eventually, after flora surveys and lots of red tape, the WA Government entered into a long-term lease agreement with the University of Western Australia that gave us access to a large, flat, sandy site called Wallingup Plain, 80 kilometres north of Perth. It had space for a future large-scale gravitational wave detector. This was dubbed the Australian International Gravitational Observatory.

In a national collaboration over the next few years, the Australian Consortium developed the Gingin High Optical Power Research Centre, with a view to developing technology for the next generation of detectors – Advanced LIGO – after the LIGO observatories had finished their development and observations.

Meanwhile I had made a commitment to create an education centre – the Gravity Discovery Centre. We would create a centre that would allow the public to participate in what we were sure would be something momentous – the opening of a brand-new spectrum.

At the time, the UWA group was collaborating closely with the French–Italian Virgo gravitational wave observatory, which was being built at Cascina, just outside Pisa. Five of our PhD students did six-month stints helping to design Virgo. I spent many weekends wandering in Pisa, writing book chapters on the back steps of the cathedral and on the lawns near the Leaning Tower, where people get photographed holding up the tower. I looked at the swinging candelabras that Galileo had timed, and I discussed with physicists about whether Galileo had really dropped weights from the Leaning Tower.

Suddenly it came clear in my mind. The Gravity Discovery Centre should emulate the wonderful Piazza dei Miracoli, which includes the cathedral, the domed baptistry and the Leaning Tower. Could such a crazy idea ever be possible?

Well, it did actually happen. I will explain how it happened below, but first I want to explain how the international context suddenly changed around this time, which allowed the Gingin Centre and all of the Australian groups to become a real part of the international quest to detect Einstein's waves.

Transforming gravitational waves from laboratory science to big science

To understand how gravitational waves were discovered, we need to understand *big science*, which began in a completely different field of physics. During the 1930s physics was focused on understanding the subatomic world. To investigate atoms, nuclei and their constituents, physicists had devised machines in which particles such as electrons or protons were accelerated to the highest possible speed. This could be achieved by making beams of particles continuously accelerate in a circular path using magnetic fields. However, this was tricky because the particles got heavier and heavier the faster they travelled (as Einstein's relativity theory had predicted), making it more difficult to force them to follow circles. The particles lost energy by emitting photons of radio, light or X-rays whenever they were deflected by the magnetic fields that were essential for making them go around in circles.

The only way to see smaller components of nuclei was for the particles in the beams to have more momentum, because the waviness or fuzziness gets smaller with increasing momentum. Electrons at low speed are huge fuzzy things. At speeds of kilometres per second their wavy fuzziness has shrunk to the size of atoms, but they have to go very close to light speed to be small enough to probe protons and the things inside them.

The problem is that more momentum means more energy loss on the corners. The only solution is to reduce the bending, which means making the circular paths bigger and bigger. Soon the machines used to do this, called cyclotrons and synchrotrons, outgrew university laboratories and were too big for individual universities to build them.

These were the machines of nuclear science. They had huge political impact because they gave us the physics of the atom bomb and, later, the even more powerful hydrogen bomb, but they also allowed us to probe the smallest building blocks of the universe.

An uncomfortable symbiosis emerged between two different forces. One force was the armaments industry, who foresaw the possibility of new and even more powerful weapons. It was fed by the Cold War between the USSR (now Russia) and the Western world led by the USA. The other force consisted of scientists horrified by nuclear weapons but wanting to uncover the secrets of the universe and make their science benefit humanity. Both wanted the same enormous machines, but for completely different reasons.

From the 1940s, bigger and bigger machines were constructed, culminating in the Large Hadron Collider at CERN, the European Centre for Nuclear Research. The construction and operation of each of these huge machines was a big science project, too big for single universities, and even single countries. Thousands of scientists were needed. CERN was a model for international collaboration. It started with 12 member states in 1954 and grew to 23. The Large Hadron Collider is the largest machine ever constructed, 29 kilometres in circumference.

The big projects were so big that they were economically significant, creating new industries and spin-offs, like the world wide web, which was invented at CERN, as well as machines and materials for cancer treatments. To work effectively, these big science projects needed a special culture, and a special sort of organisation, all based on enormous well-organised collaborations.

By the mid-1980s four small-scale interferometers had yielded promising results, and it was clear that gravitational wave detection needed to be as enormous as the particle physics machines, but it was less obvious that they needed enormous teams.

LIGO began as a collaboration between Caltech and MIT that was agreed in 1986. In 1989 a joint funding proposal for LIGO was submitted by a team that included an appointed director from Caltech, Robbie Vogt, gravitational wave pioneer, Ron Drever, who had moved to Caltech from Glasgow, and Rainer Weiss from MIT, who had triggered the German program and built a small interferometer at MIT.

Outsiders looking at LIGO at the time foresaw trouble. The two groups and their leaders were very different. Drever was intuitive, eccentric and disorganized. Weiss was focused, organised and analytic. Vogt was a tough disciplinarian with a military manner reminiscent of a sergeant major or a platoon commander. How could these three strong-minded people work together?

Despite personality difficulties, the project moved forward and survived scrutiny from review committees and antagonism from astronomers who saw the project taking money away from their pet projects. Then there was the selection of two observatory sites that involved significant pork-barrel politics. All this was achieved successfully, but tensions grew, and in 1992 Vogt dismissed Drever from the project. By 1994 the project was ready to begin construction.

The LIGO project was always planned to occur in two stages. The first stage would use simpler technology. It would prove the concept of building a huge interferometer, the possibility of building enormous vacuum pipes 4 kilometres long, and the ability to shine laser light back and forth between perfectly spherical-shaped concave mirrors, but with a radius of curvature of about 2 kilometres. This was something never done before.

The LIGO detectors were designed to be able to detect the only gravitational wave sources that we could be certain about. These were the coalescing neutron stars that Freeman Dyson had first pre-dicted. The first stage of LIGO would only be able to detect neutron stars coalescing about 50 million light years away. Based on studies of the few binary pulsars observed in our galaxy, the number of coalescences were predicted to be around one per 50 years. That means a 2% chance in detecting a signal in one whole year of opera-tion. The second stage was planned to be able to detect signals from ten times further away. Ten times further corresponds to a 1000-fold increase in volume, so signals should come 1000 times more often, corresponding to about 20 signals per year. Clearly we needed this second stage to have a good chance of detecting signals.

In 1993 something catastrophic had happened in the big-science world of particle physics. The US had been constructing a mega-sci-ence project in Texas called the Superconducting Super Collider. It would be three times the size of CERN's Large Hadron Collider. They had already spent $4 billion and constructed 14 kilometres of the planned 89-kilometre circular tunnel. But in 1993 the project was cancelled. This was terrible news for many particle physicists who lost their jobs and their big project, but suddenly there were many unemployed physicists who truly understood big science. All of them were looking for new directions.

Then in 1994 gravitational wave physicists all over the world were surprised to receive an abrupt email from Robbie Vogt: 'I have been fired'. Caltech had replaced Vogt with Barry Barish, a Caltech big-science particle physicist who had been involved in the Super Collider.

Barish had an enormous impact. He transformed the project into a true big-science project. In 1997 he created the LIGO Scientific Collaboration, which brought Australia into LIGO and much, much more. By internationalising the project, he allowed scientists from all over the world to contribute. Thanks to Barry Barish, 56 Australian scientists shared the discovery of gravitational waves – but that would not happen for another twenty years.

It is difficult to tell the story of gravitational waves in a continuous sequence because there are so many overlapping components.

By jumping forward past the discovery, I can tell the story of Barry Barish in the words of Adrian Cho of *Science* magazine, in the excerpt opposite.

It was the year 2016. Gravitational waves had been discovered and physicists were speculating who would get the Nobel Prize. Most thought it would go to the three physicists who conceived of LIGO: Rainer Weiss, Ronald Drever and Kip Thorne. But some influential physicists, including previous Nobel laureates, said the prize should include Barry Barish. However, Nobel rules prohibit a four-way sharing. This presented a conundrum for the committee. Perhaps it was that conundrum that made them decide not to award the prize for the gravitational wave discovery that year.

The next year the Nobel committee's task was simplified, because sadly, Drever died in 2017. They awarded the prize to Barish, Weiss and Thorne. The following article, written in 2016, describes how Barish transformed LIGO and why he deserved the prize.

Barry Barish

Barish, a particle physicist at Caltech, didn't invent LIGO, the Laser Interferometer Gravitational-Wave Observatory. But he made it happen. The hardware at LIGO's two observatories in Hanford, Washington, and Livingston, Louisiana; the structure of the collaboration; even the big-science character of gravitational wave research—all were moulded by Barish. 'Without him there would have been no discovery,' says Sheldon Glashow, a Nobel Prize—winning theorist at Boston University.

When Barish took over as the second director of LIGO in 1994, he inherited a project that was 'dead in the water,' says Richard Isaacson, the National Science Foundation's program director for gravitational physics from 1973 to 2002. But with Weiss, Drever, and Thorne, a theorist, tripping over one another, the project remained larval. By 1987, the National Science Foundation wanted a single director for the project, and Caltech appointed Rochus 'Robbie' Vogt.

Decisive but volatile, Vogt pulled the team together to write a coherent proposal for twin interferometers with 4-kilometre-long arms. He won crucial support in Washington, D.C. and in 1990, the National Science Board approved construction of the observatory, priced at $250 million—the biggest thing the National Science Foundation had ever attempted.

But Vogt disdained bureaucratic oversight and annoyed officials. He kept the LIGO team unworkably small and refused to supply detailed work plans or progress reports. Things got so bad that in 1993 the NSF asked Congress to hold back $43 million that the agency had requested for LIGO the following year. By the end of the year, Caltech eased Vogt out of leadership.

Barish brought swift and sweeping changes. Lanky, soft-spoken, and even-tempered, he was born in Omaha, Nebraska, and raised in Los Angeles, California, where he attended public schools. He already had experience with big collaborations, having worked in a group of 140 physicists searching for particles called magnetic monopoles at Italy's underground Gran Sasso National Laboratory. He had also worked on the biggest of big-science projects, the $10 billion Superconducting Super Collider in Waxahachie, Texas, which Congress cancelled mid-construction in 1993.

First off, Barish reorganized LIGO management, expanding the team and delegating authority. Within months, he developed the detailed work plan that the NSF wanted. Whereas Vogt stressed unfettered innovation, Barish likened LIGO to building a bridge—one that would be very long, complicated, and expensive. He revised the project to improve infrastructure, such as the vacuum chambers that hold the interferometers. He also established permanent scientific staff at the two LIGO outposts and a steady R&D program for future upgrades. These changes required a budget boost to $292 million.

Barish and his deputy, Gary Sanders shook up the culture of the collaboration. Caltech physicists had a prototype interferometer and relied on experienced individuals to restart it daily. Barish and Sanders prodded them to run it steadily 24 hours a day and to study it methodically, eliminating what Sanders calls the 'guru mentality.' The factory-like approach galled many of LIGO's leading lights, who felt devalued and quit.

Meanwhile, Barish expanded LIGO beyond the bounds of Caltech and MIT. In 1997, he brought in new expertise by establishing the independent LIGO Scientific Collaboration (LSC), the group of external scientists who would use LIGO. 'I don't think we would have made the discovery in the time we did without that massive accumulation of intellect,' says David Berley, the NSF's project manager for LIGO from 1992 to 2000. Sanders says the decision to create the LIGO Scientific Collaboration was initially unpopular among physicists who feared that big science was taking over their field. It was.

Key technical aspects of the LIGO interferometers reflect Barish's touch, too. He decided to change the lasers that pump light into the instruments from ones that squeeze light from argon gas to more powerful and reliable solid-state lasers, then just coming to market. He also pushed to switch from analog to digital controls.

Construction of LIGO finished in 1999, and it began taking data 3 years later. Barish stepped down in 2005.

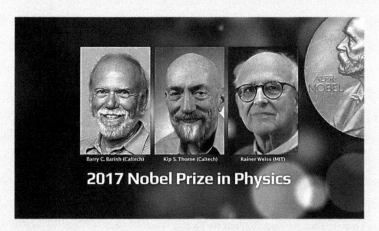

The founders and builders of LIGO, Barry Barish, Kip Thorne and Rainer Weiss.

We just jumped forward 20 years. Now we need to jump back again to uncover the immense and heroic scientific struggle to build LIGO. It makes the effort of the Wallal expedition look trivial.

Learning to drive laser interferometers: Agonising learning curve

The first step for the laser detectors was to prove that the enormously complicated instruments could operate at design sensitivity. This was an agonisingly slow and difficult process. Having built what were supposed to be the most sensitive instruments ever created, and some of the most complex, it was a massive undertaking to learn how to drive them. They did not come with instruction manuals!

The first challenge was to simply align the laser beams well enough that they would shine from one end of the 4-kilometre-long vacuum pipes to the mirrors at the other end without hitting the sides of the vacuum pipe. Then the mirrors, suspended by fine wires, had to be oriented with incredible precision so the beams could bounce back and forth between the two mirrors. Seismic vibrations were always making the heavy mirrors swing, and this had to be stopped. Once aligned, the reflected beam had to be *locked*.

Locking was the condition needed for the light to build up between the mirrors. It could only be met if the distance between the mirrors was constant to within a tiny fraction of the one-micron wavelength of the laser light – that meant positioning had to be accurate within tiny fractions of a micrometre. When the light did build up, its radiation pressure exerted an increasing push on the mirrors, which had to be compensated for using magnetic forces acting on tiny magnets attached to the mirrors.

Another problem was that the arms of the detector expanded and contracted with the tides. Most of us imagine that tides are only in the ocean, but they are also in the Earth itself. The Earth bulges on both sides due to the gravity of the moon as the two bodies swing around each other like a pair of dancers. But the Earth is spinning once a day, so its bulge moves around with its spinning. The result is that we all go up and down about a metre twice every day. This causes all distances, like the distances between cities, to expand and contract tidally. Normally such changes would be too small to measure, but for LIGO it is a huge effect. The distances between the LIGO mirrors stretch and shrink by almost a tenth of a millimetre, which is about 10,000 laser wavelengths. This would be quite catastrophic if not corrected for.

I have given you just a taste of the problems that needed to be solved. It took an excruciating amount of time to solve all the problems because every time one problem was solved, another would show up that had been covered over by the previous effect. The big problems were the easiest to solve because they were most obvious. Every problem that was uncovered was new and often quite unexpected.

Teams of young, dedicated experimenters spent endless days and nights like forensic detectives, measuring, observing, correlating, and trying to identify cause and effect. Some problems were easily solved once their origin was understood, but others needed enormous effort. These young scientists were the unsung heroes of gravitational wave detection. They were all driven by optimism about detecting gravitational waves, hoping for a discovery that would not happen for more than a decade.

Outside observers could not understand the degree of difficulty and were impatient for progress. A significant group of traditional astronomers had lost the battle to prevent 'astronomy's money being wasted on an impossible physics experiment'. That phrase was actually stated in public by a distinguished optical observatory director who shall remain anonymous, but the view was reflected in a *Nature* magazine news report in 2002, and similar stories appeared in other science media.

On the muddy floodplain of central Louisiana, two concrete tubes [actually stainless steel!] four kilometres long cut a giant 'L' into an expansive logging plantation. Inside the tubes, laser light bounces back and forth between mirrors, creating a measuring device so sensitive it can pick up the rumble of traffic entering the site, and the vibrations as nearby trees are felled.

But as evening falls and the logging stops, preparations continue for a far more ambitious piece of detective work. By the end of this year, the Louisiana detector will be looking for gravity waves – the faint ripples in space and time generated by colliding black holes and exploding stars. If detected, the ripples will provide new tests of Einstein's theories of gravity, and a way of peering into unseen areas of the Universe.

But that's a big if. The crashing trees and passing trucks – not to mention minor earthquakes – are hampering the final development of the detector. And because theoretical physicists are not sure what the waves look like, no one knows exactly what to expect. Leaders of the project – the Laser Interferometer Gravitational-Wave Observatory (LIGO) – admit that the Louisiana device, and its companion at Hanford in Washington state, might not initially spot anything. But, given that the detectors have a total cost of US$296 million, some researchers are wondering how they ever got built.

Geoff Brumfiel, *Nature*, Vol. 417, pp. 482–4 (2002)

Besides the incorrect description of the pipes, the above story contains a major falsehood: by this time physicists had an extremely good idea of the waveforms they would be looking for. These were the chirps made when a pair of neutron stars or a pair of black holes spiral together. The signal sounds like a 'whoop', rising in frequency and rising in amplitude over seconds or minutes, depending on the masses of the objects. Confidence in eventual detection was partly because known signals can be much more easily extracted from noise than unknown signals. This is what allows us to recognise complex sounds like speech in very noisy environments: we can match the patterns against the known templates stored in our neural networks.

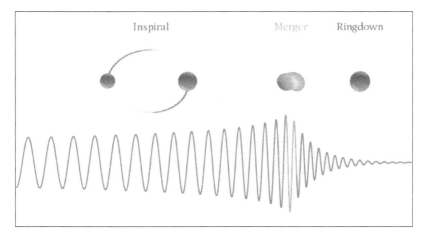

Inspiral   Merger   Ringdown

A typical gravitational wave signal traces out the spiralling motion of the two black holes. They merge together in a crescendo, and then the signal dies away as the newly formed black hole settles down like a vibrating jelly.

There was so much focus on day-to-day problems and small bits of progress that most LIGO scientists ignored the criticisms; and besides, year after year, the noise curves were improving, and inexorably they were reaching *design sensitivity* – the predicted sensitivity in the original instrument design. This was the time to start observing. This meant long-term operation of the two LIGO instruments, and the search for coincidences between them.

There were plenty of interruptions in operations when lasers failed, earthquakes disrupted operations and computers failed, but after two years, the two instruments had accumulated one year of coincident operation – i.e. an accumulated year of data when both detectors were operating at high sensitivity. As this was happening, data analysis teams were busy analysing the data. Some were searching for sudden bursts of gravitational waves from possible supernova explosions. Others were looking for continuous pure tones of gravitational waves that might be made by rapidly spinning neutron stars. Several groups were looking for coalescence signals, those chirps created as black holes and neutron stars coalesced.

In 2007 the collaboration celebrated the completion of LIGO's *science run* with a big cake that displayed the sensitivity achieved. LIGO had now formally met its commitment to achieve one year of data at target sensitivity.

The article opposite describes a strange and sad-funny event that was the culmination of these observations. By chance, it happened on Pi Day in 2011, Einstein's 132nd birthday.

LIGO's celebration cake: two detectors had been 'on the air' for a whole year, with sensitivity (the graph on the cake) matching the sensitivity that had been promised many years earlier.

2011: Big Dog, or how we thought we had detected gravitational waves but hadn't!

Up until late last year three enormous laser interferometer gravitational wave detectors in the northern hemisphere had been operating for several years at unprecedented sensitivity, listening for Einstein's elusive waves. We were searching for gravitational wave signals created by matter in its most extreme and exotic form — neutron stars, black holes and the big bang itself. The detectors were the most sensitive instruments ever created, able to detect fractional changes in spacetime geometry at the level of parts in 10^{23}, corresponding to the measurement of energy changes of less than 10^{-31} Joules per Hertz of bandwidth. Despite this extraordinary achievement, the sensitivity was still about 10-times below the level where we could be confident of detecting frequent predicted signals. For example, the mean time between loud enough chirping whistles, made when pairs of neutron stars spiral together to create a black hole was likely to be once every 50 years, so that in a year of operation the chance of detection was only about 2%.

Despite this pessimistic prognosis, many of the 1000-odd physicists in the worldwide collaborations that had created and operated these detectors remained optimistic that nature might be kind enough to provide a first signal. Optimism was high enough that the detection teams had put in place a system to alert optical telescopes to slew to the part of the sky corresponding to the most likely location of the source — determined from the arrival time of the signal in each detector.

On 16 September 2010 a coincident signal appeared in the Laser Interferometer Gravitational Observatory detectors in the USA. These 4-kilometre-long laser interferometers are spaced 2000 kilometres apart from Washington State to Louisiana. My postgraduate student Sunil Sunsmithian was in the control room at the Louisiana observatory acting as a 'science monitor'. His job was to inspect the incoming data and to alert a designated senior scientist in the event of a significant signal. This particular signal was immediately recognised as a significant event, especially after it was also identified in the data of the Virgo detector in Italy. By triangulation of the arrival times, it was determined to have come from the direction of the constellation Canis Major, the Big Dog. Within minutes telescopes in Australia, France and Chile and the Swift satellite in orbit had been automatically alerted and many images of the region of the sky were taken. The signal appeared to have come from a coalescing pair of black holes at a distance of order 100 million light years. The absence of an optical signal was not surprising. The signal corresponded to the power of 10^{23} times the power of the sun, as bright as all the stars in the universe, but all of this energy would have been emitted in gravitational waves and would be invisible to electromagnetic astronomy. Only in the case of coalescing neutron stars would we have expected a flash of gamma rays and light, a bit like a supernova.

Members of the collaborations were soon alerted, but we were all required to keep the event secret until two things had happened. First, the data needed to have been fully analysed and all possible ways that the signal could be a false positive needed to be considered. Second, we had to wait until the *blind injection* envelope was opened.

David Reitze, spokesperson for the collaboration reminded us that we had all agreed to the process of blind injections. In this process rare events *might* be injected into the detectors at unknown times by a secret team, to test the ability of the detectors to distinguish between real signals and accidental glitches in the data.

It took many months to complete the data analysis. More than 100 scientists were involved in the data analysis and almost 1000 more were following it intensely. Groups poured over data channels searching for possible electrical, optical or acoustic interference. Different data analysis search algorithms were tested and compared. Eventually it was determined that the probability that the event was accidental was one in 7000 years. Then a paper announcing the first discovery of gravitational waves was written. Finally on March 14, 2011, which happened to be Einstein's birthday, the collaboration met in Arcadia, just down the road from California Institute of Technology. Here the blind injection envelope was to be opened.

A large lecture hall equipped with 6 data projectors was packed and collaborating scientists from all over the planet were present via internet connection. First, a series of presentations gave the scientific case for discovery of gravitational waves from a coalescing pair of black holes. One talk considered whether the event could have occurred through any natural processes such as radio noise from the ionosphere. People took bets on the event being real. Most people agreed that it was 99% certain to be an injection, and yet the tension and the suspense was palpable. A leading member of the collaboration, Stan Whitcomb declared that he was sure the event was real. Champagne glasses were brought out, and filled. Finally a tiny thumb drive was handed to Jay Marx, director of LIGO. He plugged it in. A presentation appeared on all 6 screens titled 'The Envelope'.

One click later and we had the answer: it was a blind injection! A coalescence signal waveform had been injected into all three detectors, precisely timed and calibrated to appear to be the real thing.

Amid the sighs and groans and sudden hubbub, we all drank our champagne, partly to drown our sorrows, but really because this was still a triumph. The effort had proved that these enormous complex instruments could detect single rare events, and determine their nature. Rather than toasting to the discovery of gravitational waves, we toasted to Advanced LIGO, the detector that we all felt certain would mark the beginning of gravitational wave astronomy in about 2015, along with Advanced Virgo, LCGT in Japan, and hopefully LIGO-Australia.

D. Blair, published in *The Conversation*.

The above article was written in March 2011. The predicted detection of gravitational waves in 2015 was indeed sustained. Seven years after the first detection described in the next chapter, LCGT in Japan has been re-named KAGRA and is still being developed, while the hoped-for LIGO-Australia is now the planned LIGO-India. It may start construction soon.

In Australia, like in the US, the one-pot mentality was alive and well. The idea is *if they get fed, we will starve.* Gravitational waves were seen as a threat to conventional astronomy. Under strong pressure from astronomers, the Australian Government refused the generous $150 million offer from LIGO and the National Science Foundation to build a third detector in Australia, much to the disappointment of the Australian gravitational wave community.

The last leg in the search for gravitational waves had begun years before the Big Dog fiasco. Australia had been closely involved since we joined the LIGO Scientific Collaboration. Like all big science, participants were all closely linked through many working groups. Each group had members from across the world, and each focused on specific issues and challenges, such as improved isolation against vibrations, or improved optical coatings for the mirrors, or improved techniques for analysing data.

Chapter 10

The road to discovery

The story of gravitational wave discovery is the story of Advanced LIGO. From the outset it had been planned as the second step for laser interferometers, where the very best available technology would give a ten-fold increase in sensitivity. Advanced LIGO was already being planned while LIGO was coming into operation. Barry Barish had internationalised LIGO with the LIGO Scientific Collaboration, and led by David McClelland at ANU Australia was admitted as a full partner in Advanced LIGO.

The Advanced LIGO project was a partnership between USA, Britain, Germany and Australia. Australia provided certain key components and technologies, including technology for measuring distortions in the light waves passing through the mirrors, aligning the output beams and preventing the detectors from becoming unstable, as well as supercomputer-based data analysis systems to dig signals out of the noise.

Advanced LIGO is a collaboration of about 1000 physicists across the world. Every university and research institute provided state-of-the-art know-how, expertise and equipment, all assembled into two 4 kilometre by 4 kilometre laser interferometers and data processing centres. The total search cost more than $1 billion, but has developed know-how, technologies and training that are certain to be worth far more than this to our future.

To explore Australia's role, we have to step back in time to the 1970s and 80s.

Carlton Caves and squeezing the vacuum

In the 1970s, Vladimir Braginsky was recognising that gravitational wave detection would be limited by the quantum limits to measurement, and realising the need for enormous optical power if interferometers were to be sensitive enough to detect gravitational waves.

Braginsky introduced us to a new term: *quantum non-demolition*. This awkward Russian term describes a powerful concept. It recognises that all measurements act back on the system being measured, but if you are clever you can engineer the back action so instead of acting back on the thing you care about measuring, you act back on something you don't care about.

Imagine you are trying to measure the length of a table with a tape measure, but your hand is very wobbly. If you could make sure your wobbles were always sideways, you could still read the tape measure quite precisely, but if your wobbles were lengthwise, your measurement would be very uncertain.

In 1927 German physicist Werner Heisenberg had formulated quantum uncertainty in the *Heisenberg uncertainty principle*. In its most famous form it says that in any measurement, the uncertainty in position times the uncertainty in momentum (i.e. speed times mass) must always be bigger than the tiny quantum constant called Planck's constant.

This was part of a more general concept where quantities you might measure can be put together in pairs (like position and momentum) and all the pairs will satisfy similar relations. If you measure one thing very precisely, you always make the other one less precise.

Imagine you were trying to measure the position of a hanging ball using your tape measure and your wobbly hands. You might manage to get a good measurement of the position at one instant, but immediately your wobbly hand will have pushed the ball and it will swing around with a new uncertainty in its position. This is quantum uncertainty – the size of the wobble is set by the quantum constant, but it is mixed between two variables, in this case the position and the momentum.

We saw something very similar in Chapter 9, when we talked about measuring mirrors with laser light and realising that if there were too few photons there would be a high statistical uncertainty, but if there were too many photons their momentum would push the mirrors around. This was an example of the uncertainty principle: try to measure position too precisely with lots of photons and the momentum of the mirrors would become uncertain.

In 1981 Carlton Caves, a young researcher at Caltech, published a paper in which he analysed the problem of uncertainty in laser interferometers and came up with an astonishing conclusion. In his paper he was asking: where does the uncertainty come from, and can it be reduced?

'This paper presents an analysis of the two types of quantum-mechanical noise, and it proposes a new technique – the 'squeezed-state' technique – that allows one to decrease the photon-counting error while increasing the radiation-pressure error, or vice versa.

'The key requirement of the squeezed-state technique is that the state of the light entering the interferometer's normally unused input port must be not the vacuum, as in a standard interferometer, but rather a 'squeezed state' – a state whose uncertainties…are unequal [in the two quantum variables of the light].'

Caves was saying that uncertainty is something that leaks into the interferometer from the outside world. Physicists call it the *vacuum*, or the *quantum vacuum*. It is what you have if you have no photons at all – like total darkness. But total darkness is not *totally* dark. It is full of *uncertainty photons* and they can enter every interferometer. These, Caves said, were the cause of uncertainty. These uncertainty photons are normally called *vacuum fluctuations* or *zero-point photons*. Quantum physics tells us that when you take away all of the photons, there is always something left behind, which has the energy of half a photon. These half photons are everywhere, and you cannot make anything that is free of them. These are the ones I call uncertainty photons.

Caves pointed out that you could modify the incoming uncertainty photons with a device that today is called a *squeezer*. It uses entangled pairs of photons to create a special type of light amplifier. Caves was suggesting that this amplifier would modify, not real light, but vacuum fluctuations – those uncertainty photons.

The special kind of amplifier you need to do this magic works in the same way that children 'pump' a playground swing. On a swing,

Carlton Caves, a young researcher at Caltech in the 1970's.

you have to time your pulling exactly right to make the swinging get bigger. In more technical words, you have to pull at the correct *phase* of the swinging. If you don't, the swinging will not build up. This same phase dependence of the light amplifier allows it to modify the wave pattern of the light that passes through it.

Caves concluded his paper like this:

'The squeezed-state technique outlined in this paper will not be easy to implement. A refuge from criticism...can be found by retreating behind the position that the entire task of detecting gravitational radiation is exceedingly difficult.

'Difficult or not, the squeezed-state technique might turn out at some stage to be the only way to improve the sensitivity of interferometers designed to detect gravitational waves. As inter-ferometers are made longer, their strain sensitivity will eventually be limited by the photon-counting error...Further improvements in sensitivity would then await an increase in laser power or imple-mentation of the squeezed-state technique. Experimenters might then be forced to learn how to very gently squeeze the vacuum before it can contaminate the light in their interferometers.'

Caves said that you just needed to modify the invisible beams that get into an interferometer. I found it very difficult to believe that the idea of these invisible beams of 'vacuum photons' could be anything other than something mathematical. Was this *really* how quantum uncertainty arose? Eventually I was convinced in a simple and dramatic way, which I will describe later.

Caves idea of squeezing started an experimental campaign to verify his theory. But as Caves had predicted, it was difficult, and it did not go well. Four years later at a laser optics conference, a well-known laser optician Marc Levinson reported:

'...[squeezed] states have eluded experimental demonstration, at least so far. From an experimentalist's point of view, squeezed state research can be best described as a series of difficulties that must somehow be overcome. What follows in the proceedings are nine sections, titled 'First Difficulty' all the way up to 'Ninth Difficulty', nothing more nothing less.'

Soon after that gloomy report came the first glimmer of hope, and then a few years later in 1987, Ling-An Wu, a PhD student in the lab of Caltech professor Jeff Kimble, was the first person to observe practical squeezing. She wrote 'A degree of squeezing of approximately fivefold is inferred...An explicit demonstration of the Heisenberg uncertainty principle...is made from the measurements.'

Thirty years later, when she was a scientist at the Chinese Academy of Sciences, Ling-An Wu visited Perth and charmed the Einstein-First team with accounts of those pioneering experiments as well as the 3000-year history of optics in China, all recounted in her impeccable Oxford English.

In the years after the first demonstrations of squeezing there was not much progress, but as gravitational wave detector projects progressed, the need for squeezing became more and more apparent

Ling-An Wu, the first person to demonstrate quantum squeezing, on top of the Leaning Tower of Gingin looking across the flat gravitational wave observatory site.

as the detector builders confronted difficulties with the very high laser power they needed.

Gravitational wave researchers at the Australian National University led by David McClelland had led efforts to transform squeezing from laboratory experiments to practical tools for enhancing gravitational wave detectors. Gradually teams around the world got better and better at making squeezing sufficiently noise free. By 2013 McClelland's team were able to write: 'The field of squeezed states for gravitational-wave (GW) detector enhancement is rapidly maturing…Squeezing is included in the baseline designs of all future interferometer upgrades…and squeezed state generators can now produce the magnitude of squeezing likely to be needed over the next decade.'

Vacuum fluctuations are everywhere, empty space is alive with them, they are real, and they can be squeezed!

While squeezing clearly worked, I still found the concept very difficult to imagine. Visiting McClelland's labs at ANU one day I was shown a squeezer in action. It was a massively complex device with laser beams criss-crossing all over a huge table and a computer screen showing the background noise. A student showed me where the squeezed states were entering. I asked her 'If I were to block the beam with my finger, the squeezed vacuum would be replaced by ordinary vacuum…right?' She nodded. 'Then it will get noisier, right?' 'Yes', she replied. 'So my finger test will prove that squeezed vacuum is real.'

She let me do the finger test and the noise level got worse. Somewhat bemused, I now had to accept that Carlton Caves's explanation was all true. Vacuum fluctuations are everywhere, empty space is alive with them, they are real, and they can be squeezed! They contaminate all measurements and are responsible for quantum uncertainty in interferometers.

The Australian Consortium team with smiles after they won the 2020 Prime Minister's Prize for Science for their contribution to the discovery of gravitational waves. In these COVID times we were unable to celebrate together.

Controlling thermal deformation of mirrors

Squeezing offered a way of overcoming some of the difficulties with laser power, because the effect of squeezing was to change the statistical fluctuations of the laser light so it had less photon counting noise, but this was at the expense of the forces acting back on the mirrors. These forces act most strongly at low frequencies like the low notes on a double bass, but at these low frequencies many other technical problems like residual seismic vibrations were far more serious.

One of the biggest difficulties for Advanced LIGO had always been anticipated. This was a phenomenon called thermal lensing, which is not unlike the mirages you often see when you look along a hot road and see the sky shimmering in the road. The light is bent as it passes through thinner heated layers of air, so instead of seeing the road, you see the sky.

If you send a powerful laser beam through a flat piece of glass, one photon in every million may be absorbed. For the best fused silica glass this may only be one milliwatt out of every kilowatt of laser power, but this still makes the glass get warmer, expand and develop a tiny bulge – just like the bulge of a magnifying glass, but a million times weaker.

But when laser beams have to travel a few kilometres to a distant mirror, this is enough to cause quite strong focussing. The Australian Consortium decided to investigate this problem. By this stage the University of Adelaide had joined the collaboration. It was led by Jesper Munch, a Danish expert in very powerful lasers, and Peter Veitch, who had been a crucial member of the team at UWA that built the Niobé gravitational wave detector. Peter had joined the Glasgow group as a postdoc before returning to Adelaide, now with expertise in both the bar detectors and laser interferometers. He proposed a way we could make our 80-meter-long test arms at Gingin extra sensitive to thermal lensing. This would allow us to develop methods of compensating for the thermal lenses by adding heat around the edges where there was no laser heating, to flatten out the thermal bulges.

There was one catch: the absorption of laser light is not uniform, nor is it predictable. There needed to be a way of measuring and mapping the thermal expansion, but this needed enormous precision. The Adelaide team developed sensors that shine thousands of tiny parallel beams through the huge glass plates. A camera images the dots, and a computer searches for small changes in the spot positions.

The Adelaide technique is very reminiscent of the Wallal expedition, where the positions of a hundred-odd beams of starlight were used to map the gravitational lensing of the Sun. The distortion of the pattern of the laser spots allows a thermal map of the mirrors to be created. I remember the first movie of thermal distortion in a big mirror at Gingin: you could see an initially flat profile growing into a huge mountain after the laser was turned on. That huge mountain was probably one tenth of one nanometer high!

The Adelaide technology is now used in all the detectors for monitoring their mirrors. It has found hot spots in some of the multimillion dollar mirrors which, despite the most stringent quality control, still manage to have imperfections that make life very difficult for the detector builders. It remains a crucial technology for testing detectors and diagnosing their problems.

Preventing instability

In 2005 our team had predicted that the Advanced LIGO detectors would also suffer parametric instability, which could be an intractable problem because there were so many vibrational harmonics of the mirrors that could start to ring. The only place parametric instability had been observed was on the niobium bar – a very different, and much simpler situation. We needed to test our theory, but the problem was that the experimental conditions needed for observing instability were as extreme as the conditions needed to make a high sensitivity gravitational wave detector – we needed superb vibration isolation, huge laser power, and mirrors with near perfect ringing properties, and we also needed the very large laser-beam spots you would have if the laser beams were travelling 4 kilometre to their mirrors. Advanced LIGO was doing most of this with their multi-million-dollar budgets, but we had to do it cheaply.

We thought it might be possible if we used innovative home-grown technology. Our PhD student John Winterflood had discovered many new high-performance vibration isolation techniques. He helped a team of PhD students to turn the ideas into a 3-metre-high vibration isolation system with, we believed, better performance than the vastly bigger and more expensive systems created by LIGO and Virgo. I used to call it the 'quietest place in the universe'!

Next, we needed a way of suspending mirrors so they would ring almost perfectly. Here we used a legacy from the niobium bar technology. Ju Li and I invented this new suspension, using little suspension structures made of niobium and hooked into little holes in the mirrors. Another PhD student proved their performance. Zhao created a mirror design that enabled huge fat laser beams to resonate between the suspended mirrors. He designed it so that the light would build up 3000 times, much more than in LIGO, so that we could get enough laser power to compensate for our small size of only 80 metres.

About this time my son Carl, who had built a portable eye-camera for astronauts to use on the International Space Station (where it is still used today), asked Zhao if he could do a PhD in gravitational waves. From then on gravitational waves became a family business with me, my wife Li Ju and my son! Meals around the dinner table were intense and technical!

Carl joined the experimental team just at the time that most of the home grown technologies had come together, and he began the ordeal of testing and searching for parametric instability. Our struggle reminded me of LIGO's struggle to get their first detectors operating correctly. Again, there was no instruction manual. We had to learn how to drive our wobbly machine full of intense laser light that was powerful enough to move the mirrors into all sorts of unwanted swinging. Three or four PhD students were the master drivers.

Intricate experiments are like young babies with colic. You can't leave them, they scream unpredictably, and they need constant attention. Usually they settle down late at night. But when everything starts to work properly you can't sleep; settling down is the time you can begin to take data.

Wherever intricate experiments are taking place, the lights are on most nights. Knowing this, at Gingin we built a set of bedrooms very close to the lab. Signs warn people not to disturb the sleeping researchers. It was like this when we built the Niobé detector. Boxes of biscuits and cups of coffee kept us going, and we made friends with the night security staff. Those were the times when the eureka moments happened.

It took us from 2005 to 2014 to prove that our parametric instability predictions were correct. Every step of the way had been difficult, but we found solutions, and every phase had its own eureka moments. It took 13 PhD projects to get there.

The final eureka moment came when Carl saw steadily rising vibrations in one of the mirrors.

For the whole team this was a big breakthrough too, because now we knew we had mastered almost all the technologies needed to build a full-scale gravitational wave detector. Then we would also be in a position to get support for our ultimate dream: to build a full-scale gravitational wave detector at Gingin.

In 2014, just as we were submitting a paper on the first observation of parametric instability in a suspended cavity, we heard that the commissioning team at LIGO were observing just the effect we had predicted. The mirrors were steadily ringing up, quite slowly, but after some time becoming unmanageable, causing the whole control system to fail. They could not turn up the laser power to the level needed: the greater the laser power, the faster the ringing increased. At Gingin we had struggled to make instability happen because our 80-metre arms, lower power laser and smaller mirrors made it much more difficult. At LIGO it was too easy.

Fortunately we had already done many experiments where we had artificially induced instability and then tested different methods of stopping it. By 2014 we had quite a large tool kit for instability repairs! Carl Blair immediately offered to help, and before long he was working with the commissioners at Livingston, Louisiana, bringing all the skills we had learnt to help solve the problem.

In December 2014 we received a heart-warming letter from Matt Evans, an MIT LIGO scientist who was working on the commissioning of the new detectors, and, as he explained, writing a paper reporting on the observations of parametric instability in LIGO.

'Hi David and Zhao,

'For the exceptional contributions you and Zhao have made to understanding parametric instability, and the sustained effort both of you have made to clarify the phenomenon in modeling and experiment, I would like to put your names on the author list of this paper. This is also because I feel that you have been uniquely key to pushing Slawek and the rest of us to keep an eye on this phenomenon, with the result that we were prepared when it was observed.'

The letter was especially welcome because for years we had been going to conferences and reporting on the predicted problem and sometimes receiving cold and negative responses to our predictions. We were always bearers of bad news, and people wanted to shoot the messengers. After Evans's letter we felt redeemed and appreciated at last.

Meanwhile Carl threw himself into the job of understanding the instability in LIGO and developing control strategies. Parametric instability can be tuned by shaping the mirrors. This can be done by warming them up with heaters. But the mirrors warmed up spontaneously. The art of control was to carefully add heat to the mirrors, but it was very tricky.

Working with the LIGO commissioning team, Carl developed control strategies and codes for controlling the heating and monitoring the ringing signals that gradually allowed LIGO's laser power to be increased. That took until 13 September 2015. The first observation run was due to start, and all systems were running. The real eureka moment arrived the next day.

John de Laeter and the Gravity Discovery Centre

John de Laeter

John de Laeter was a remarkable physicist who had an enormous impact on the West Australian contribution to the discovery of gravitational waves, and indirectly to the entire Australian contribution to gravitational wave astronomy. He had started life as a science teacher, but a science teachers' conference he attended changed his career.

'I heard two of the world's experts battling it out on how the universe began — the Big Bang Theory versus Steady State Cosmology. It inspired me and I decided there and then to go back to university and do a PhD in physics and get involved in these astrophysical questions.'

De Laeter went on to do distinguished research in geo-chronology (the dating of minerals and meteorites by their isotopic composition) while he was the inaugural head of physics at the institute that became Curtin University. His research led to commercial instruments for microanalysis and he made enormous contributions to the study of West Australian geology and mineral exploration.

De Laeter's passion for science education never left him, and he put enormous effort into establishing the Science and Mathematics Education Centre at Curtin University, the Scitech Discovery Centre in Perth and the Gravity Discovery Centre at the gravitational wave observatory site near Gingin. He also made a huge contribution behind the scenes to establish the site for the Square Kilometre Array radio telescope.

The gravitational wave community remembers John de Laeter's enormous efforts in raising private-sector donations for the creation of the Gravity Discovery Centre. This in turn supported the scientists' case for research funding, because this was a tangible demonstration of public support for science research and a way of communicating science directly to the public. The Gravity Discovery Centre supported the case for the creation of the Australian International Gravitational Research Centre as a West Australian centre of excellence in 2005. This in turn helped in the creation of the national centre of excellence for gravitational wave discovery, OzGrav, in 2015.

De Laeter was active as chair of the Gravity Discovery Centre Foundation from its formation in the year 2000, up until his death in 2010. The first construction had been the Southern Cross Cosmos Centre, (now called Gingin Observatory) containing a cluster of telescopes for public viewing. In a space of less than ten years the Gravity Discovery Centre grew to include its main Discovery Gallery building, the Cosmology Gallery — an enormous buckyball dome that celebrates our quest to understand our place in the universe — a 1-kilometre scale model of the solar system and the iconic Leaning Tower of Gingin, where children play Galileo while learning that their 222-step climb was a climb against time, speeding up the passage of time as predicted by Einstein.

From the Leaning Tower of Gingin you can see the huge flat plain reserved for a future gravitational wave observatory, and, in the foreground, the High Optical Power Research Facility with its 80-metre interferometer arms, as well as the Geosciences Australia magnetic observatory, which benefits from the quiet isolation of the site to monitor changes in the magnetic field of the Earth that protects the Earth from space weather and flares on the Sun.

Chapter 11

Detetected at last

Exploring the universe in gravitational waves

Gravitational waves are akin to sound waves that travel at the speed of light through empty space. The first detections in 2015 mark the birth of a brand-new type of astronomy – gravitational wave astronomy. We have started to explore a brand-new spectrum. Like deaf people switching on their bionic ears for the first time, humanity has now turned on its gravitational ears, and now we can listen to the sounds of the universe.

Gravitational wave signals

Long before gravitational waves were detected people had calculated exactly what the signals would be like. Imagine two black holes orbiting each other. If we were looking at their orbit edge on, it would look like two black holes crossing back and forth in line, but if the orbit was face on, we would see a circular motion. The stretchings and shrinkings of space that we would observe would be different in each case because of our viewing angle.

One detector alone cannot discern this effect, but if you have detectors with different orientations (which is unavoidable if they are widely spaced on the round Earth) they can tell the orientation of the black hole binary system in the sky from the relative size of the signal in each detector. This is one way that gravitational wave signals avoid observation angle ambiguities that are present in other forms of astronomy.

If two black holes are orbiting each other they give out gravitational wave energy. This makes them spin around each other faster and faster, so the gravitational wave frequency rises over time. If one black hole is enormous, the frequency will be low because the orbit takes a long time, even at light speed. The highest frequency of gravitational waves that two black holes can create depends on the size of the combined black hole: this size depends only on their total mass. How the gravitational wave frequency changes with time tells you how much energy is being given out: it changes faster if the two black holes are equal in mass and more slowly if they are unequal. The loudness of the signal depends on how far away it is.

The bottom line is this: gravitational wave signals completely define the system being observed – in other words, they tell us everything we need to know about it.

'Everything' is not a lot, because black holes are amazingly simple things – they say black holes have no hair! What we can determine from every signal is its distance, the mass of each black hole, the shape of their orbit and how they are spinning. That is enough to allow much detective work to be done to find out how and where they were formed.

2015: The first signals

On 14 September 2015, just as Advanced LIGO was being readied to begin long-term observations, a signal appeared almost simult-aneously at both LIGO observatories 2000 kilometres apart. The signal at first seemed too good to be true, and for many months mundane explanations were sought. Could it be hacking? Could it be lightning? Could it be a computer glitch or accidentally coinci-dent vibrations from outside getting into both detectors?

The waveforms from binary black holes are highly specific. They rise in frequency and rise in amplitude as the black holes get closer together. The loudness of the signal tells the distance of the source, the frequency and how it changes with time tell the masses of the two black holes, and the final ringing tells the spin and mass of the final black hole.

The information content of binary black hole signals is astonish-ing to conventional astronomers, who have to contend with the extreme difficulty in disentangling brightness and distance: is a star bright because it is close or because it is very powerful? For gravitational waves, it is as if every signal comes with a packing slip that specifies the mass of each object, the mass of the final object, their distance in the universe and even how they are spinning.

The accuracy of the determination depends on the strength of the signal compared with background noise from the detector. In the case of the first signal, there was noise, but taking the noise into account, there was also clear agreement that the signals in each LIGO detector came from the same source. The wave came in obliquely, causing the signal at Livingston, Louisiana, to arrive 8 milliseconds earlier than the signal at Hanford, Washington State. If the time difference had been more than 11 milliseconds, it would have been ruled out because its speed would have had to be less than light speed.

The first detected gravitational wave signal came from a pair of black holes spiralling together and merging at a distance of about one billion light years. The black holes had masses of 29 and

36 times the mass of the Sun. This in itself tells us a story about stars in the universe, but like a good detective story, there are several options that we will explore below.

First, let's think a bit more about the signal. From the distance of the black holes, we know that the signal we observed began its journey more than one billion years ago, when life on Earth had not progressed past single-cell organisms. The signal was like a tsunami of space. It expanded out like the ripples on a pond when you throw in a stone. But remember, this is a spherical expansion, so it is perhaps more like a fireworks explosion in the sky.

The gravitational wave tsunami got weaker and weaker as it expanded further, but it was never weak. It contained the energy of three Suns converted into gravitational energy in a few tenths of a second. This energy was spread across its entire spherical wavefront.

The gravitational wave burst had completed more than half its journey to our detectors before the first multicellular organisms appear in the fossil record. It had completed about 95% of its journey to Earth when a giant asteroid wiped out all the large dinosaurs on Earth. Still the signal had a long way to go. As it passed the Milky Way's nearest large neighbouring galaxy, Andromeda, early hominids were walking the plains of Africa. It reached the edge of the Milky Way about the time our Homo Sapiens ancestors were spreading from Africa to Europe, Asia and Australia. When the tsunami was closer than the nearest star, the LIGO detectors were being assembled, and it was well within the cloud of comets that encircle the solar system while the detectors were being fine-tuned. The tsunami finally arrived when the detectors were running and collecting data, but still some hours before official data recording was supposed to begin!

When the signal did arrive, it oscillated the distance between Perth and Sydney by about the size of a proton, and the mirrors in the detector were moved by about one thousandth of this amount.

The waves reveal the final few cycles of the merging black holes, beginning when the holes were spinning around each other 25 times every second. Within a tenth of a second the two holes merged into one. The new hole was born vibrating wildly, but over a few hundredths of a second the vibrations damped down. A single black hole now continues its journey. Far in the distant future, it may meet another black hole and eventually repeat the merger process.

Where did these black holes come from? We presume that they were born from enormous stars made from the primordial gasses hydrogen and helium when the universe was very young. They may have been born as a pair of giant stars that remained together after each star collapsed in turn and became a black hole. This might have happened in the very early universe – say, 13 billion years ago – or more recently. It all depends how close those black holes were to each other when they were formed.

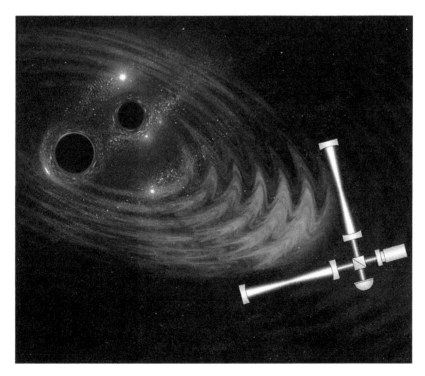

This visualisation shows the gravitational waves created by a black hole merger sweeping through the universe and eventually encountering a laser interferometer capable of measuring its passage. IMAGE COURTESY CHRIS MESSENGER

There is a very narrow range of spacing allowed for those two black holes we saw merging. If they had been born further apart, they would merge so slowly that they would never make it to coalescence in the age of the universe. And if they were born a bit closer together, they would have coalesced too quickly.

Ours was just right. It coalesced a bit more than 1 billion years ago, and the waves rippled towards us ever since, to pass through the solar system on 14 September 2015. For every pair of black holes that we see coalescing, there must be vastly more of them that already coalesced or will never get there in the age of the universe.

How many is that? We have no idea, but with statistics we can work it out. That means we need to detect lots more signals.

The first detection: a student's perspective

Carl Blair

I was working at the LIGO Livingston Gravitational Wave Detector in 2015. The subject of my PhD — a strange interaction between light and matter that can make mirrors vibrate — had suddenly become very relevant. When the Advanced LIGO detectors had been turned on in 2014, the mirrors vibrated making it impossible to detect gravitational waves. So I was there to stop the mirrors vibrating. We had demonstrated the phenomena and methods to suppress the vibrations in experiments at the Gingin gravity science facility.

The effort to get the detector working was phenomenal, and it was phenomenally exciting. Scientists were visiting from all over the world to install their technology or to just lend a hand. A small commissioning team was hard at work making the detector work and searching for all the little (and big) sources of noise that prevented the detector from reaching design sensitivity. Detector engineers were hard at work making their subsystems failsafe. And commissioners and detector engineers were training the operators to keep the instrument running. Some of the commissioning work was easy and rewarding, and some of it was impossibly tedious. The return on work was measured in detector sensitivity — if it improved, that was a triumph; if it didn't, you had to go back and try something different. At some point we received the edict 'enough is enough — stop fiddling with the detector and start collecting some science data!' The two detectors had reached a respectable sensitivity, able to detect neutron stars colliding 200 million light-years away. Well, that is what we estimated.

In the first weeks of September 2015, the calibration team had not finished the final calibrations of the Livingston detector. The calibration allows us to be sure that a mirror moving one thousandth the diameter of a proton is actually measuring this distance. (That is how sensitive the detectors were!). I had been working in the control room late at night on my project. I had been keeping my ears open and learning how the calibration was done. On the night of 14 September the final calibration was complete, confirming that we could detect neutron stars colliding 200 million light years away! We all went home at about 3 am, leaving an operator in the control room to keep the detector running. The detector at Hanford was also running. One hour later, at 4 am local time, something almost unbelievable happened. The most perfect gravitational wave signal, just like the predicted signals I had seen many times, showed up in the data. That was a little awkward, because official observations had not yet started.

The signal was so perfect and so perfectly timed that I did not believe it was real. I was not the only one. For weeks the corridors of LIGO were full of questions like: Do you think it is real? How could it be such a perfect signal? Could it be a blind injection? Blind injections were a touchy subject. We all new about Big Dog, the blind injection in 2010 that had caused hundreds of scientists to spend thousands of hours vetting the signal, making sure everything was working properly, writing the paper to tell the world, and then being told that it had been a simulated injected signal just to test out the system. Understandably, many scientists were still miffed.

Once we had this signal there was a lot to do. We had to vet the signal, to make sure everything was working properly. Again, hundreds of scientists put in thousands of hours. We checked all the auxiliary data from the detector — there were 100,000 channels monitoring everything about the detectors from all the control signals and all the vibration, sound, earthquake and electrical and radio signals you can imagine. Teams worked though different subsystems. After months of this, my spirits were starting to wane. Was this all going to be another waste of time? I couldn't help but still be excited though. Adam Mullaley, a friend from Wagga Wagga who is a detector engineer at Livingston, had been telling me for months that he was in charge of the hardware injection pipeline and it was not even ready on the day of the signal. But I was not sure there was not another back-door way a signal could have been injected.

In November we had a commissioning meeting at MIT. All the top brass were there, and hundreds of collaboration members were online. It changed my view. We had a presentation by one of the MIT leaders, Matt Evans, who described the work he had done to ensure it was not a malicious injection. He described scenarios where malicious actors synchronise phones with GPS at either LIGO site to inject electronic signals into the detector to simulate a gravitational wave. He showed how, no matter where they injected the signal, the code he had developed would identify it as a malicious injection. This must have taken him a month. If he had put all this time in, maybe this was a real gravitational wave signal.

After this meeting there was a growing sense of excitement and an increased amount of work. It was serious. A paper was going to be written. With a thousand authors, we all tried to put in our 2 cents worth. A paper writing team put it all together, it was reviewed and improved in round after round, and finally the paper was published on 11 February 2016. The celebrations lasted a year, maybe two; maybe we are still celebrating. This detection marked the first time humans had heard the whispers of spacetime.

Happiness was the discovery of gravitational waves. The UWA gravitational wave team after the discovery.

Fun facts: The unity of opposites

The first detection described a gravitational wave explosion that was the most energetic event ever observed by astronomy — more than 3 solar masses of pure gravitational energy were emitted in less than 1/10th of a second.

- The power emitted in gravitational waves was 50 times the power output of all the stars in the universe.
- The burst of energy that passed through the solar system was the most powerful energy flash ever observed by astronomy.
- The actual energy detected by the gravitational wave detectors was the smallest amount of energy ever measured by any sort of instrument.
- The movement of the mirrors in the detectors was the smallest distance change ever measured.

Suddenly

Black holes explained many astronomical phenomena,
but never had been directly observed. The discovery
of gravitational waves was momentous.

Suddenly everything changed.

Suddenly we went from hoping and searching to knowing.

Suddenly we knew that Feynman had revealed the truth.

Suddenly we knew that the theory of detection was correct.

Suddenly we went from theory and prediction to observation.

Suddenly we could hear the vibrating
event horizon of a black hole.

Suddenly we knew that the black holes of
Einstein's theory really exist in the universe.

Suddenly we knew with certainty that space is populated
by black holes, some in the form of binary pairs slowly
spiralling together, others alone and silent.

Suddenly we knew the sensitivity needed
for listening to the entire universe.

Suddenly we could read binary black holes like time
capsules to learn about our stellar ancestors.

Suddenly we knew that the spacetime we inhabit
is constantly rippling like an ocean surface.

Suddenly new understandings were within our
reach that hitherto had been merely dreams.

That was the meaning of discovery.

David Blair

Chapter 12

How to make gold: The first neutron star coalescence

About ten times in the last 2000 years, supernovae have astonished and frightened the world. They are brilliant spectacles. They happen fast.

Astronomers have unravelled the story of what actually happens, and the Crab supernova described in Chapter 6 was a key part of unravelling this story. Stars maintain their size by pressure. In normal hot stars like the Sun, pressure is created from the momentum of very hot electrons, nuclei and photons pushing outwards in just the same way that atoms continually bombarding the inside surface of a balloon push it outwards. The momentum, which depends on the mass and speed of the particles in the star, comes from the energy released in the fusion reactions, when the nuclei of atoms like hydrogen fuse together to make heavier atomic nuclei. But fusion ceases once the growing nuclei have been have built up to element number 26, iron. Iron is the most stable nucleus. It is like the ashes of a campfire – the stuff that is left when the wood has all burnt. Once the core of the star has become iron, there is no more heating, and the pressure reduces. This makes the core of the star shrink, and this increases the gravity, which squeezes it still further. Could anything stop it from collapsing?

For seven decades astronomy books have proclaimed that all the heavy elements like gold, platinum and uranium are made in supernova explosions. This turns out to be false.

When I described the discovery of neutron stars in Chapter 6, I avoided a very important part of the story, because I wanted to keep it for this chapter. It explains how a gravitational wave discovery in 2017 solved the mystery of where gold is made in the universe.

The story takes us back to 1930, when a 19-year-old Indian student called Chandrasekhar was on a boat trip from Calcutta to England to study astrophysics with Eddington at Cambridge. On the long voyage he applied the new theory of quantum mechanics to the mysterious stars called *white dwarfs* – tiny stars, 100 times smaller than most stars and about the size of the Earth. They are not small because they are light-weight. They have as much mass as the Sun, a million times the mass of the Earth. Astronomers assumed that all stars must end up like this, but no theory had explained them.

Chandra, as he was called, realised that if you squeezed matter a million-fold, the electrons around atoms would be squeezed out of their orderly clouds to wander randomly throughout the star. But would it be random?

In all atoms, electrons arrange themselves in energy levels because electrons refuse to share the same energy state. This is called the *Pauli exclusion principle*. The electrons have to arrange themselves in a sort of *energy staircase*, starting at the bottom and filling upwards. Protons and neutrons also follow this same rule.

Chandrasekhar did something seemingly outrageous. He applied the concept of the energy staircase of electrons in atoms to the vast scale of a dying star. He realised that Einstein's relativity would have big effects, just as it does in atoms of gold. In gold atoms, which have 79 electrons, the fastest electrons become a bit heavier because of $E = mc^2$, and this change causes gold to have its distinctive colour. Treating the star as a giant atom, Chandra showed that near the top of the staircase the electrons would be approaching light speed. The heavier the star, the faster the electrons would travel.

Next, Chandra showed that the increasing electron mass near light speed would make the star shrink as you added more matter. Add still more and the star collapses. He predicted that no white dwarf could be heavier than 1.44 times the mass of the Sun. Up to the critical mass, the star behaves like a single vast atom, but afterwards it must collapse.

Chandra's supervisor Eddington tried to suppress Chandra's work, stating, 'I think there should be a law of Nature to prevent a star from behaving in this absurd way!', but Chandra eventually got it published and 50 years later he was awarded the Nobel Prize! By this time thousands of white dwarfs had been studied and their masses were always less than his predicted critical mass.

A year after Chandra's work, neutrons were discovered, and Zwicky made his prophecy about the connection between neutron stars and supernova explosions. Oppenheimer had followed Chandra's arguments but replaced electrons with neutrons. A star made of neutrons also had to obey the non-sharing rule that forces the neutrons to arrange themselves into an energy staircase with one step for each neutron. Because neutrons are nearly 2000 times heavier than electrons, they have 2000 times more momentum. This lets them exert 2000 times more pressure. So, roughly, a neutron star could be 2000 times smaller than a white dwarf, but have just as much mass. Its density would be trillions of tonnes per litre! But, like a white dwarf, if more mass were added it would shrink, and once the fastest particles got very close to light speed, the star would collapse.

Supernova 1987A

In January 1987 a star exploded in Australia's backyard, a small galaxy called the Large Magellanic Cloud just 150,000 light-years away. Being beyond our own galaxy, it was only just visible to the naked eye, but through telescopes its form was quite magical, and almost every telescope on Earth examined its light. Rings of gas were brightly illuminated as ejected material smashed into them, and the abundant light allowed careful analysis. There was plenty of evidence for radioactive iron and nickel, and even barium and strontium, but no sign of the really heavy elements like gold. The message: supernovae do not make gold.

The lack of gold in supernova 1987A did not seem to bother many astronomers. The official line remained in all the textbooks: heavy

Supernova 1987A in the Large Magellanic Cloud. This was the first supernova visible to the naked eye since Kepler's supernova of 1604 and the discovery of the telescope. The rings were created by blast waves expanding into space.

metals are made in supernova explosions, but a few astrophysicists were very concerned.

In 1989 Tsvi Piran and co-workers started to think carefully about what would happen when two neutron stars coalesced. Remember, this is neutron matter. It is only able to exist under the huge gravitational pressure needed to squeeze electrons and protons together, because quantum uncertainty wants them to be far apart. Individual neutrons are unstable radioactive particles that break up into protons and electrons. Atomic nuclei with too many neutrons always radioactively decay.

You can see what happens when two neutron stars get really close together in the image below. The stars distort to become tadpole-shaped, and long tails of material fly into space, driven out by the rapid spinning, while the heads merge.

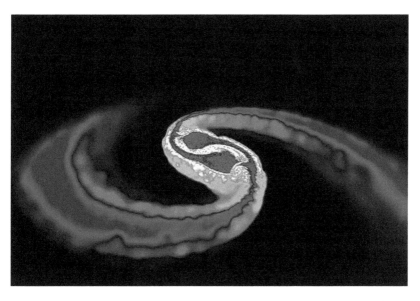

Computer simulation of two neutron stars coalescing. The cores merge but long tails of material are thrown into space. The colours are not real: at the million-degree temperatures of neutron stars, most of the light is in the form of X-rays.

The matter of the neutron stars is as dense as an atomic nucleus, but at the facing surfaces of the two stars, the neutrons are suddenly free of the immense gravitational pressure that created them. Now radioactive fission begins. The de-pressurised neutron matter breaks up into smaller and smaller fragments. Neutrons convert to protons by emitting electrons. The huge density of neutrons plus newly formed protons makes extremely unstable oversize nuclei. They decay almost instantly. Within 1–2 seconds the nearly pure neutron matter has converted into heavy nuclei. Most are unstable and continue to decay until they reach a stable configuration that might be gold or platinum or uranium.

The neutron matter explodes like a gigantic atom bomb, creating a cloud of highly radioactive heavy elements that all decay according to their particular half-lives, adding extra energy to the exploding mass of material. Astrophysicists call this a *kilonova*.

Atom bombs make tiny amounts of gold when a few kilograms of uranium or plutonium spontaneously explode. The kilonova is an explosion of up to 100,000 Earth masses of neutron matter. Some of that mass becomes energy due to mass difference between the initial nuclear matter (mostly neutrons) and the final heavy elements left behind. This is a tiny fraction. About 16,000 Earth masses are thought to be heavy elements, and almost all of this explodes out into space. This includes perhaps five Earth masses worth of gold, and the same for platinum and other heavy elements.

The Earth is 4.6 billion years old, but our galaxy is at least 10 billion years old. Long before the Earth was born binary neutron stars were coalescing and throwing heavy elements into space. This mixed into huge gas clouds that collapsed to make new stars and planets. This way gold was mixed into the fabric of planet Earth – very rare, but very beautiful.

Given that one ten-billionth of the mass of the Earth is gold, one coalescence makes enough gold for 50 billion planet Earths! You don't need too many neutron star coalescences to provide the galaxy with plenty of gold to go around!

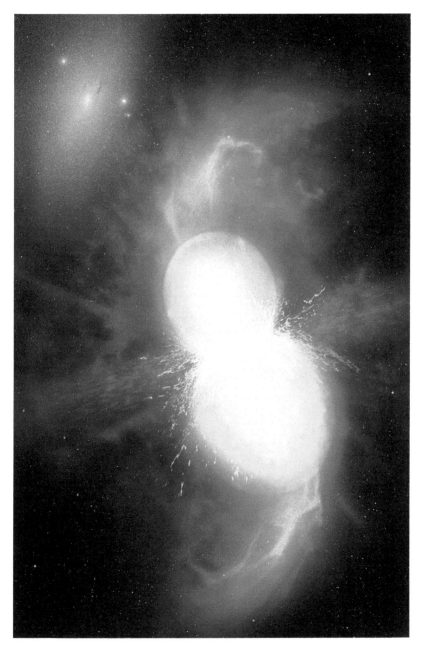

Artist's conception of a kilonova, with neutron-rich matter spewing into space.

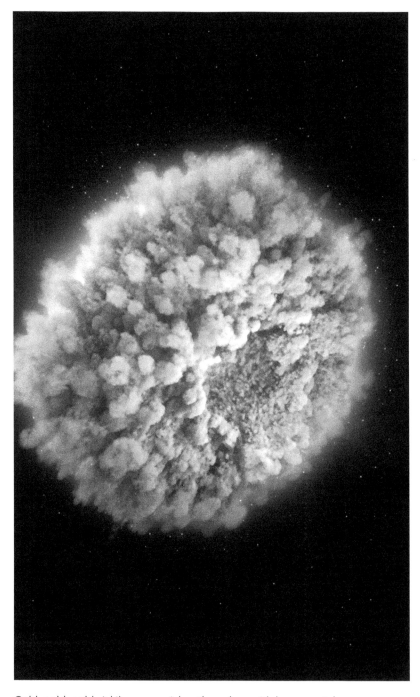

Gold, gold, gold. A kilonova enriches the galaxy with heavy metals.

That is a rough sketch of the theory developed by Piran and others. The theory was in place but where was the evidence?

It all came 18 years later, on 17 August 2017, when the LIGO and Virgo gravitational wave detectors detected the unmistakable slow siren sound of two neutron stars spiralling together over the space of a minute. The long slow chirp told us the mass of the two stars. Clearly, they were neutron stars. About 1.7 seconds later a powerful burst of gamma rays was detected by the Fermi gamma-ray space telescope.

The first visible light ever seen from a neutron star coalescence, which showed the characteristic signature of heavy elements like gold. This photo was taken by the Swope Telescope, named after Henrietta Swope, who helped determine the size of the Milky Way and the distances to other galaxies.

The gravitational wave data and the gamma-ray data gave a rough location in the sky. Optical astronomers sprang into action, and within hours the Swope telescope in Chile had identified a new star in a distant galaxy called NGC4993 that is about 130 million light years away. This was the light of a kilonova explosion. The Zadko telescope at Gingin and ANU's Skymapper telescope imaged the kilonova and tracked it as its brightness rapidly decayed. The decay was not like that of a supernova, which dims over 100 days or so. The light was more like the light of a giant atom bomb and the radioactive decay of newborn heavy elements. It also showed the characteristic infrared colours created by heavy elements.

Suddenly we knew where gold is created in the universe.

The gravitational wave event told us other things too. The light had taken 130 million years to get here, and so too had the gravitational waves. They arrived within 1.7 seconds of each other – 1.7 seconds in a 130-million-year race. Their speeds were identical to within a few parts in 1000-trillion. As predicted by Einstein 101 years earlier, gravity travels at the speed of light. It was amazing that the very first measurement of the speed of gravitational waves should yield such a spectacularly accurate answer.

The last big thing from this event was that it was the first time we had been able to identify the host galaxy of a gravitational wave event. Remembering that gravitational wave signals carry information about their distance in the universe, we could now use this distance along with the recession speed of the host galaxy (which is routinely measured for galaxies by looking at the *red shift* of their light), to get the first independent measurement of the expansion rate of the universe, called the Hubble constant. Amazingly, this measurement depends on only one thing: the strength of gravity (called the universal constant of gravitation), which has been measured in laboratories for hundreds of years.

With just one signal, the first measurement of the expansion speed of the universe was not conclusive, but it proved the method, and with more signals in the future, gravitational wave signals will

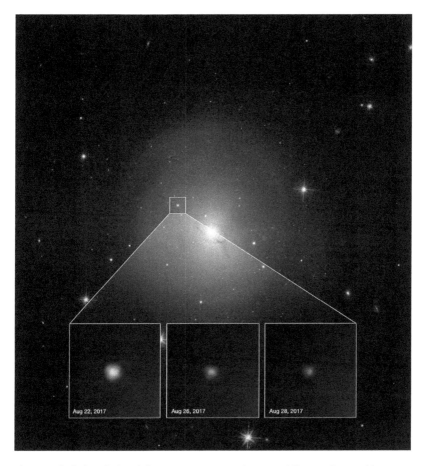

The rapidly fading light of the neutron star coalescence kilonova imaged by the Hubble Space Telescope a few days after the event. The kilonova still outshone the light of the billions of stars in the host galaxy NGC4993, which are individually undetectable, providing only a diffuse foggy glow. The other bright objects are nearby foreground stars in the Milky Way or distant galaxies in the background.(NASA IMAGE)

allow us to measure the universe much more precisely than we ever could before.

There was one thing we did not see in the gravitational waves: that was the moment when the two neutron stars formed a black hole. In fact, we lost all hearing of the event while the stars were still apart, going around each other 100 times per second. The really interesting part – when they reached 1000 orbits per second and became a jiggling mass of nuclear matter – was missing, because the detectors are very hard of hearing at high frequency. There were just not enough photons in the gravitational wave detectors to resolve the last moments of the life.

The problem reminded us once again of the need for high laser power in our detectors. We wished for better detectors to solve the new mysteries that now seemed within reach. The next part of this book will tell you how we plan to solve this.

Before you embark on this, if you would like to make gold like the old alchemists tried so hard to do, I now have the perfect recipe for you: see it in the box opposite!

How to make gold

Preparation time: 5 billion years

For this recipe you will need two stars, each about 15 times the mass of the Sun.

1. Set the stars spinning around each other.

2. Watch carefully until you have seen both stars run out of fuel and collapse in supernova explosions; first one, then the other.

3. Now you should have an orbiting pair of neutron stars. Make sure that they are close enough together that they emit a significant amount of gravitational waves.

4. Watch carefully as the stars get closer and spin faster.

5. Once the two stars have reached near contact, look out for the neutron matter being stripped away and flung outwards.

6. Watch the neutron matter blobs as they break up and explode like an atom bomb.

7. Use a fine sieve to collect up the gold before it is spread across the Milky Way.

8. Carefully store your gold for use in future planets. You should expect to collect several Earth masses of pure gold.

9. No cleaning up necessary! The double-mass neutron star will disappear from the visible universe, leaving behind just a gravitational wave signature.

10. Not only have you made gold, but you have also witnessed a mini-big—bang happening in reverse.

11. Study those waves to help you understand how the universe was born.

Chapter 13

Discovering the graveyard

Neutron stars for dinner and first sight of a black hole

Gravitational wave discoveries accelerated in the years following the first detections as LIGO and Virgo were improved. The KAGRA detector in Japan joined the collaboration, further increasing the international nature of gravitational astronomy, even though their detector was not yet sensitive enough to detect signals. The main technical improvements were focused on increasing the laser power and introducing squeezing to reduce the photon statistical noise as Carlton Caves had proposed almost 40 years earlier.

Improving sensitivity has a huge pay-off because every time you double the sensitivity to mirror motion, you become sensitive to signals twice as far away. Twice the distance means the volume of the universe you are now sensitive to is eight times larger, which means on average eight times as many signals.

By 2021 the total number of signals detected was approaching 100, with many more anticipated in the next long observing run of the detectors in 2023. The diagram on page 255 shows the tally of events in time sequence, from the very first discovery of two 30 solar mass black holes merging in 2015 to the first neutron star merger in 2017 and now many more. Each line shows the masses of the two coalescing objects, and the final outcome, measured in solar masses. We call it the stellar graveyard plot.

The biggest tombstones in the stellar graveyard are black holes between 80 and 200 solar masses. Most of them were created by a pair of monster black holes, but one of them was already big, and gobbled up a mere 20-solar-mass companion.

The existence of the near-100-solar-mass black holes is mysterious. Stars this massive have not been seen with telescopes, and astrophysics theory suggests that stars this massive should be unstable and obliterate themselves in giant explosions. Perhaps the existence of these monsters tells us that these black holes have already gone through a cycle of coalescences, building up step by step from lower mass black holes. For this to be true, there must be plenty of places in the universe where black holes can accumulate – a sort of dating service for black holes where lonely singles can find partners.

There are two main locations where this might happen. One is in the dense cores of globular clusters, like the beautiful 47 Tucanae cluster in the southern sky shown in the image opposite. Black holes are thought to slowly migrate to the centre of such clusters, where they may meet up with a partner and begin their final dance of death. Another possibility is that they meet up with partners in spinning disks of stars around supermassive black holes in the centres of galaxies, like the region around the 4-million-solar-mass black hole at the centre of our Milky Way that was imaged in 2022.

The biggest black hole coalescence up to 2022 is summarised in the LIGO Scientific Collaboration's infographic and observed waveforms on page 256. Its name, GW190521, is code for its discovery date. This pair of black holes emitted 8 solar masses of gravitational wave energy. The huge masses of these black holes meant that the peak frequency was about 60 Hz – a really low brief hum. This made it detectable at the enormous distance of 17 billion light-years.

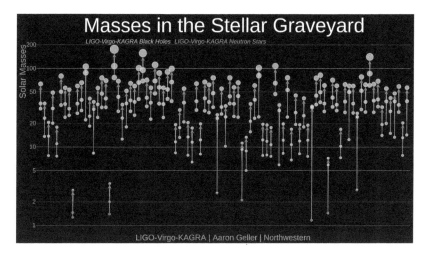

The stellar graveyard is full of black holes, each a tombstone for the partners that coalesced and disappeared from the universe leaving only a black gravitational dimple to mark their presence.

Globular clusters like 47 Tucanae in the southern skies contain hundreds of thousands of closely packed stars. Their central cores are likely places where black holes meet and begin their dance of death as the move towards coalescence.

'Hang on', you say, 'that exceeds the age of the universe!' In an expanding universe we are seeing the most distant parts when the universe was young and much smaller than it is today. The quoted distance takes into account the expansion of the universe and the fact that gravitational wave energy is lost in the expansion. Distances no longer match up to ages.

The signals in the three detectors are shown here: because of the huge masses, all that the detectors could clearly hear was the ringing of the newly formed black hole that weighed in at 142 solar masses.

Returning to the graveyard plot, we see plenty of tombstones for more-normal-sized black holes – say, six to ten times the mass of the Sun. These are likely to have been made from stars that were born together as a spinning pair of giant stars, that then burnt up their fuel and collapsed into black holes.

Some other fascinating tombstones in the graveyard tell us a story of black holes eating up neutron stars. Neutron stars are the only stars small enough that a black hole can swallow them whole.

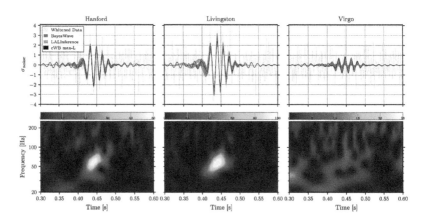

The heaviest pair of black holes created a monster. The signals show up as waveforms, the briefest of hums in the three detectors, each with a different strength because of their orientations, and fractionally displaced in time as the signals arrived successively at each detector.

GW190521

The most massive black hole collision observed so far

Discovery

21 May 2019

Distance

17 billion
light years away

3 Detectors

Three detectors made the observation: the two LIGO detectors in the USA and Virgo in Italy.

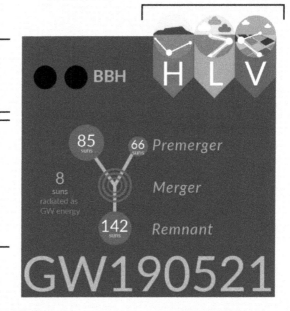

H L V

Binary Black Hole Merger

BBH

High Masses

This is the heaviest pair of black holes which have ever been observed colliding.

85 suns

66 suns — Premerger

8 suns radiated as GW energy

Merger

142 suns — Remnant

GW190521

Origin Story

The black holes which collided to make GW190521 are so massive that we're not sure how they were formed.
One option is that they are both the result of previous black hole collisions.

Ringdown

The black hole formed in the collision continues to vibrate after the merger, and "rings" like a bell for a while. This lets us test our theories.

Once again Einstein's *General Relativity* passed this test.

All larger stars, like Cygnus X-1, which we met in Chapter 6, are much too big for a black hole to swallow. A 10-solar-mass black hole is 30 kilometres in diameter, while a 10-solar-mass *star* is millions of kilometres in diameter – it is like a mosquito trying to swallow an elephant! All the tiny black hole can do is strip off gas into a swirling disc that slowly gets sucked down the plug hole.

The neutron stars stand out in the graveyard plot because neutron stars must always be less than their critical mass, which is roughly twice the mass of the Sun. Each of them can be seen at the bottom of the graveyard plot, and each is partnered with a black hole from six to twenty times the mass of the Sun. These events are almost certainly black hole feeding events, but so far only one has been seen with telescopes.

In a few cases the graveyard plot shows objects in the 2–3 solar mass range. It is quite uncertain if these are black holes or neutron stars. It will take much more data to disentangle these mysteries.

As soon as the first signal was detected, people began to ask how often black hole coalescences happen in the whole universe. Remembering that each of the events shown in the graveyard plot has a known distance, you can turn that distance into a volume and ask what fraction of the universe we were listening to. The numbers are still not precise, but the gravitational wave observations tell us that across the visible universe, black holes like those in the grave-yard plot are coalescing once every few minutes, while neutron stars coalesce a few times a minute.

Assuming that we are not living in a special moment of time, we can be certain that over the age of the universe, space has been steadily filling with more and more ripples, each of which carries up to about 5% of the mass of the black holes that created them. Assuming that this process has been going on for billions of years, the total mass of LIGO-type black holes could add up to the mass of 10,000 Milky Way galaxies (the Milky Way has a mass of 10 trillion Suns), while the total mass of the gravitational waves they created could be 100 Milky Ways.

The most crucially important number we need to know for understanding black holes in the universe is the most difficult one to determine – how many black holes are out there that don't alert us to their presence by conveniently coalescing, either because they are alone, are too close to a partner (have already coalesced) or are too far apart (will never coalesce). If the observed ones were one in a million, the total mass of black holes would be enormous – maybe enough to explain the mystery of dark matter. But if they were one in a 100, the consequences would be less drastic.

We must wait for answers to these questions as more signals are detected, but meanwhile we can be content with the revelations so far. Wallal proved that space is deformed by matter, and gravitational waves tell us that space is like a rippling ocean and that these ripples represent a significant amount of the mass of the universe – not dominant but not insignificant either.

A dying universe

The gravitational waves rippling past us are recording a dying universe. Matter is disappearing into the void of black holes where space and time come to an end. The expanding universe we observe today tells us that the universe emerged in a single event where space and time began. But coalescing black holes tell us that for much of the universe, the end comes in a multitude of collapse events spread throughout space. All these little collapses happen while the rest of the universe continues to expand.

What is the role of all these black holes? Not just those we hear coalescing, but also those others we infer to be out there because their coalescence timing cannot be so well timed that we are hearing all of them now.

Whatever their role, one thing we can say is the coalescing black holes give us a great new mapping tool. If we assume that wherever there is matter, there are black holes merging, we can use their distribution to map out the whole history of the universe.

The universe is a history machine. Its history is inscribed in its own existence, just like the history of the Earth is inscribed in its rocks.

How do we read the history? Remember that the signals we hear in black hole death spirals are signals that have travelled enormous distances through the universe. The graveyard plot records signals mostly coming from the nearest 10% of the universe, but as we improve detectors in the future, we will be able to hear the signals from 100% of the visible universe – from the earliest moments of the universe, as soon as the first binary black holes began to coalesce.

Like ancient tombs, each signal is like a time capsule from an earlier time in the universe. As we listen, we may hear very ancient time capsules, and also more recent ones. Each tells us how far back in time it is from because each signal tells us its distance. The whole history of the universe is played out in this symphony of gravitational wave chirps, rising tones happening slowly and fast, higher and lower, and some just a ping of an oscillating black hole. So far, we do not have enough signals to disentangle the symphony, but as time passes and detectors improve, the symphony may answer all our questions. When did black holes first form in the universe, what was their range of sizes, how did their sizes change over time, and did the supermassive black holes in the centres of galaxies all form from many smaller black holes?

Expanding the gravitational wave spectrum

Gravitational waves have so far revealed black holes from a few to 200 solar masses. But there should be gravitational waves created whenever supermassive black holes coalesce, and when smaller black holes and stars fall into supermassive black holes. All these signals happen much more slowly, because the frequency of gravitational waves decreases as the black hole mass increases.

Australian scientists are deeply involved in the gravitational wave astronomy for these much lower frequencies. There are two key technologies. The first turns the pulsars – those rapidly spinning neutron stars – into gravitational wave detectors. To understand how they can do this, remember that gravitational waves stretch and shrink space. Matter is tied to space by inertia, so that, when space expands, neutron stars and planet Earth respond accordingly – the distances between them expand and contract.

The laser interferometer detectors exploit the fact that gravitational waves stretch and shrink space in perpendicular directions: as a gravitational wave passes, you get shorter and fatter, then taller and thinner, back and forth at the frequency of the wave. For supermassive black holes the frequency would be about one cycle per three years, while for a star falling into the black hole in the centre of the Milky Way, the frequency might be one cycle per hour.

At the lowest frequency of about one cycle per year, you can search for changes in the arrival times of pulses from pulsars. The project aiming to do this in Australia is called the Parkes Pulsar Timing Array or PPTA. The idea is to look for timing fluctuations in pulsar signals that could be caused by passing gravitational waves. The biggest effect will always be from two pulsars perpendicular to each other. To discern this, you need to see if the pulse arrival times fluctuate more for perpendicular pairs of pulsars. So far there is a hint of a signal from statistical measurements on lots of pulsars. Many Australian scientists from CSIRO, Sydney and Swinburne University are involved in this long-term study, which so far covers 20 years of pulsar data. New telescopes like the Square Kilometre Array and China's FAST telescope will get better and better data in the future.

For gravitational waves at around a cycle per hour, a project called the Laser Interferometer Space Antenna is being developed by Europe and the USA. It will create a multimillion-kilometre laser interferometer in space during the 2030s. This detector requires exquisite engineering and three spacecraft orbiting in a triangular pattern following the Earth's orbit around the Sun. Australian scientists at ANU are deeply involved in the laser technology required for this mission.

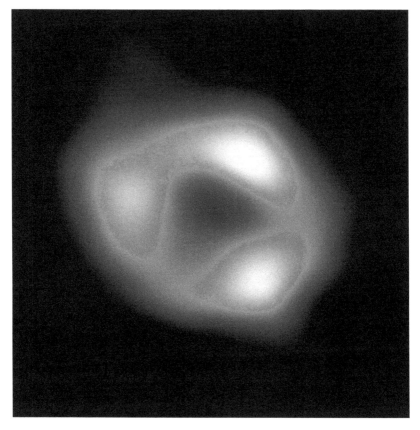

Supermassive black holes have long been inferred because of their gravitational effects and the huge jets of radiation created by matter when it is trapped and heated in accretion discs around the black holes before the matter falls inwards. In 2019 the Event Horizon radio telescope project succeeded in imaging a 6-billion-solar-mass black hole in the middle of galaxy M87. In 2022 the same team imaged the 4-million-solar-mass black hole in the centre of the Milky Way. Both of these black holes had long been inferred, but the images have given us direct confirmation of their existence.

The big questions

The first detection of gravitational waves marked the beginning of the exploration of the gravitational wave spectrum. We have found a whole population of coalescing black holes that tell us something about our dying and transient universe. It has opened up a slew of questions that we can now think about being able to answer:

— How many black holes are there in the universe?
— What is the connection between supermassive black holes and LIGO black holes?
— What fraction of the universe has turned into black holes?
— Where are the coalescing black holes we observe located: around the supermassive black holes in the cores of galaxies, in globular clusters or drifting through intergalactic space?
— How much of the baryonic (normal) matter and how much of the dark matter in the universe has been lost into black holes?
— Does the total surface area of black holes in the universe always increase, as predicted by Stephen Hawking and Jacob Beckenstein?
— Is the singularity inside a black hole always hidden as proposed by Roger Penrose, or could it emerge at the moment that two black holes turn into one?
— Can we measure dark matter and dark energy by mapping the universe in gravitational waves?
— What happens when dark matter falls into black holes?

We will need many more detections and good statistics to answer these questions. As detectors improve there are many more sources of gravitational waves to be discovered. Today we have exposed the tip of the iceberg. In the next decades the dark side of the universe will be revealed by gravitational wave astronomy.

Chapter 14

Detector technologies and the future

Gravitational wave detectors are quantum instruments, sensitive to the smallest amount of energy ever measured. One wrong photon of laser light can cause a disturbance. We have had to learn how to measure suspended masses while minimising how much we disturb them. All of this required deep changes in our understanding of measurement, combined with enormously challenging technical innovations. These are still underway.

In 1886 Heinrich Hertz discovered electromagnetic waves. When asked about their usefulness he was adamant: electro-magnetic waves were 'of no possible use whatsoever'. How wrong he was!

As a child I was fanatical about radios. I had a huge collection of ancient radios with their beautiful glass valves, huge coils, 'high tension' transformers and pulley systems for tuning their intricate capacitors. My bedroom used to be piled high with the e-waste of the 1950s and 60s, much to my mother's despair. Through doing and making and dismantling, I learnt the incredible story of radio technology and the successive innovations that took us from the first weak signals picked up on the airwaves to the solid-state transistors that were the pinnacle of technology when I was a child. My crowning success was when I succeeded in broadcasting commercial ads on one of the ABC's ad-free radio stations!

Sixty years later our radios have become our mobile phones. The technology of the 1960s is archaic, and radio technology from the 1900s is almost unrecognisable. Our phones today would be completely unrecognisable to the pioneers of radio technology like Marconi.

Will the story of harnessing the gravitational wave spectrum contain similar revolutionary changes? I dare not speculate how gravitational wave technology may look in 2122.

The story of gravitational wave discovery has been like set of parallel relay races, with the baton getting passed from one team of racers to another. While one race was coming to fruition, another race was underway. Before one race was finished, the baton was thrown to another team.

We have seen how, while theoretical physicists were trying to understand the fundamental ideas, astrophysicists were making connections with astronomical observations. Teams invented lasers while others struggled with thermal noise and seismic noise. The first struggles to create resonant bars ended in consternation, while a new team was inventing cryogenic bars and discovering the pitfalls of quantum measurement. Just behind, but rapidly overtaking, came the physicists building laser interferometers. Soon after the cryogenic bars had reached the pinnacle of their technology, laser interferometers brought a quantum leap in international cooperation. The demands of big science coalesced the field into a much more coordinated effort. All the teams began to work towards the single goal of gravitational astronomy. Today Australia is deeply embedded in that effort.

The common theme in the development of gravitational wave detectors has been the challenge and the opportunities from deeper understanding of the quantum physics of detectors. Today, many more innovations are in the pipeline and are being developed so that the detectors can take another ten-fold leap in sensitivity over the next decade.

While we have little idea how we could harness gravitational waves for communication or other purposes, the technology needed for building detectors has had enormous applications for 'useful things' like mineral exploration, quantum computing, precision sensors, seismic intrusion sensing, ultrasensitive radars, pollution monitors and vacuum technology. Across Australia, industry-funded projects exploit innovations that were forced by the extreme difficulty of detecting gravitational waves.

Gravitational wave detection and Australia

Gravitational wave detectors have intrinsically poor directional sensitivity, but excellent distance sensitivity (for coalescence signals). We have seen how signals from binary black holes and neutron stars encode their distance as well as their masses and spin rates.

To determine the direction of gravitational wave sources, many widely spaced detectors are required. There is a special need for an Australian detector because it fills the biggest gap in the worldwide array. A detector in Australia would bring a dramatic improvement to the global array, helping both to identify sources and to dig deeper into the noise to extract weaker signals.

With a southern hemisphere detector adding to the world array, we will be better able to map signals, identify the galaxies or galaxy clusters they came from, and tell radio, optical and X-ray telescopes where to look. This is essential if we are to understand where the black holes are located and where they came from, and to find the answers to many of the new questions.

The addition of a southern hemisphere detector improves the sensitivity of the world array, roughly doubling the number of accessible sources. Accidental false positives from disturbances and interference are suppressed by the *power* of the number of detectors in the array, so one extra detector multiplies the performance of the array. An Australian detector can be sufficient to reduce interference to a negligible level.

Thus, Australia has an important future role in gravitational astronomy. We have 40 years of experience, innovations and technologies, and a dynamic young team that was part of the first discovery, who are ready to build Australia's future in gravitational astronomy.

A pinnacle of technological innovation

The birth of gravitational wave astronomy marks a pinnacle of technological innovation, but it also marks the power of human cooperation. The discoveries of one person, Albert Einstein, early in the twentieth century motivated a century-long struggle for understanding. Eventually it enabled people to understand how to make quantum measurements on huge objects and how instruments can reach their ultimate sensitivity. The actual creation of these instruments could not have been achieved without massive cooperation by 1000 physicists across more than 100 institutions.

The building of gravitational wave detectors required multiple breakthroughs. People had to learn how to make mirrors precise to atomic dimensions, and to be able to reflect laser beams with less than one part in a million being lost. Others had to learn how to suppress the natural vibrations of atoms caused by heat, and vibrations caused by earthquakes, cars and walking people – vibrations that were one billion times bigger than the gravitational wave signals. Then physicists had to learn how to pick tiny signals out of a huge background of noise, and learn how to ensure that the detectors would not act back on themselves and create spurious signals from the power of the lasers that operate them. The lasers that were the hearts of the successful detectors themselves harnessed Einstein's 1917 discovery of stimulated emission, and Weber's realisation 33 years later that stimulated emission could amplify light. Thereafter, myriad discoveries by hundreds of solid-state physicists allowed near-perfect sources of laser photons to be created.

Australian scientists played a pivotal role in all these break–throughs. Fifty-six Australian scientists contributed to the discovery of gravitational waves and shared the International Breakthrough Prize. Today gravitational wave astronomy is rapidly growing across Australia and across the world as new discoveries challenge our understanding of the universe.

We are only at the very beginning of the exploration of our new gravitational wave spectrum. Gravitational waves are humanity's new sense, our sudden new ability to listen to the universe. Our detectors are bionic ears for humanity, giving us a brand-new way of knowing the universe, one that emphasises the reality that space itself is a dynamic stage that responds to every movement of matter in the universe. We are like sailors, navigating a spacetime ocean. New detectors across the planet and in space far from the Earth's gravitational disturbances will lead to new revelations. The next part of this book presents Paul Davies, who foresaw the potential revelations of gravitational wave discovery more than 40 years ago. He still looks to the future.

Part 3

FOUNDING NEMO

By Paul Davies

Epilogue

We have reached the epilogue of the story about the discovery of Einstein's new universe, the universe of curved space and gravitational waves. You have learnt about the century-long struggle to understand the meaning and implications of the general theory of relativity, and glimpsed some of the amazing revelations brought to us in gravitational waves. In this epilogue we shall recap some of that story, and leave you with dreams of future discoveries.

Gravitational waves and the secrets of the universe

Gold has long exercised a peculiar fascination for mankind. It was the dream of alchemists to transmute base metals into this precious substance, an endeavour that proved fruitless. In fact, the origin of gold remained a mystery until well into the twentieth century, when physicists came to realise that all metals, indeed most chemical elements, were made by stars. But, as related in earlier chapters, in the case of heavy metals, the details remained hazy until August 2017, when astrophysicists witnessed, in a mere fraction of a second, the creation of enough gold to outweigh the entire Earth. The source of this astonishing stellar alchemy was an event of almost unimaginable violence: the collision of two neutron stars in a faraway galaxy.

Neutron stars are the remnants of large stars that run out of fuel, resulting in their cores imploding catastrophically. They typically possess a mass of about one-and-a-half Suns, but squashed into a ball the size of a city, a density so great that even atoms are crushed by the intense gravitational force to form neutrons. Sometimes a pair of neutron stars are locked in orbit around each other, entering a death spiral that terminates in a monstrous encounter when the two objects smash together and then collapse, in an instant, into a black hole.

Spurred on by the events of 2017, a group of Australian scientists is now proposing to build a giant instrument dubbed NEMO – Neutron Star Extreme Matter Observatory – to detect the fine details of neutron star collisions. In the split second it takes for the stars to coalesce, they re-create the conditions that would have prevailed in the universe just after the big bang, so NEMO should provide clues about the very birth of the cosmos. But neutron star mergers happen so fast that studying them demands a highly specialised instrument with a $100 million price tag.

NEMO should provide clues about the very birth of the cosmos.

The key to observing these awesome encounters has been told in this book. It makes use of one of the strangest phenomena known to science: gravitational waves. The story began with Albert Einstein's theory of relativity, published in 1905, with its weird space-and-time-warping predictions. In a sweeping extension to his work, outlined in a series of lectures in Berlin in 1915, Einstein unveiled his 'general theory of relativity', regarded by many as the finest intellectual achievement of mankind.

The general theory of relativity, or general relativity for short, is a theory of gravitation. It toppled Isaac Newton's account of gravity, formulated more than 200 years previously and used by generations of astronomers to work out the orbits of planets and comets. Einstein's iconoclastic theory conceived of gravitation in a completely novel way. Newton, famously inspired by the sight of a falling apple, treated gravity as a force of attraction between material bodies that reaches across space and weakens with distance.

According to general relativity, however, gravity isn't a force at all but a distortion in the geometry of space and time. The Sun, for example, creates a space-and-time warp around it and the reason the Earth orbits the Sun along a curved path is not because the Sun pulls on our planet, as Newton described it, but because Earth follows the shortest possible path through the warped geometry.

We are familiar with material objects that bend, such as blocks of rubber. But space can bend too, and the Wallal expedition set out to prove just that. The idea of bent, or warped, space takes some getting used to; because most of us think of empty space as simply a featureless vacuum. The illusions resulting from warped space are similar to that created by fish-eye lenses and fairground mirrors. General relativity predicts a similar effect, but with the lensing done by space itself. Look at the featured photograph overleaf. It shows a spacewarp created by the enormous mass of a galaxy, which causes images of more distant objects to be sculpted into distinctive arcs. The light beams from these far-flung objects are bent as they pass through the curved space around the intervening galaxy.

The reason that space can bend is because it is elastic. Nobody noticed that before Einstein. As space is incredibly stiff, it takes an enormous mass to bend it by just a smidgeon. And space can not only bend, it can stretch, shrink, twist and buckle too. The expansion of the universe, for example, can be envisaged as the space between galaxies swelling or stretching. Every day, a hundred billion billion cubic light years of additional space appears within the observable universe. The twisting of space is also observable: Earth's rotation very slightly drags space around with it, an effect

An example of the spacewarping effect, taken with the NASA/ESA Hubble Space Telescope (depicting GAL-CLUS-022058s, located in the constellation of Fornax). The image of the more-distant galaxy is distorted by the gravity of the intervening galaxy to form a smeared-out arc.

ESA/HUBBLE & NASA, S. JHA. ACKNOWLEDGEMENT: L. SHATZ

that has actually been measured using gyroscopes carried aboard a satellite called Gravity Probe B, launched in 2004.

Given the elastic nature of space, it is no surprise that it can vibrate too. That much was predicted by Einstein already in calculations published in 1916 and 1918. He noticed that the equations of general relativity could describe undulating spacewarps that travel at the speed of light. The best way to envisage these ripples is to imagine their effect on matter. Suppose a gravitational wave were to come at you face on. The distortion of the space inside your body means you could be stretched vertically and squeezed horizontally. A moment later that would be reversed – with vertical compression and horizontal extension – as the wave passed through its cycle.

Needless to say, nobody has ever experienced the physical effects of a passing gravitational wave directly in this manner. That's because such waves' traction on matter is extremely small. Even if Jupiter smashed into Saturn, nobody on Earth would feel a gravitational wobble. Contrast this with more familiar electromagnetic waves: stand in front of a military radar antenna and you'd be fried. Electricity is far more powerful than gravity, which is why you can make a balloon stick to the ceiling simply by rubbing it.

The long journey to understanding gravitational waves has been described in this book. For a long time, it was assumed that gravitational waves would be too feeble to ever be directly observed. But, in the 1960s, a handful of visionary physicists thought it might be possible to detect gravitational ripples from exploding stars by carefully suspending a bar of metal in a vacuum chamber and looking for minute vibrations using electronic sensors. By the 1990s several such bars were in operation, including one at the University of Western Australia designed by physicist David Blair. Nothing was picked up. Ideally, the scientists wanted a supernova explosion in the Milky Way, but the last one to be observed was in 1604, so we might be in for a long wait for the next.

Meanwhile, attention shifted to another source of gravitational waves: neutron stars. Thousands are known to astronomers from the radio pulses they emit. When they occur in pairs, they orbit each other, and in so doing they should emit a steady stream of low-frequency gravitational waves, according to the theory. The waves rob orbital energy from the stars, causing them to spiral together and move faster and faster until they hurtle round each other at a substantial fraction of the speed of light. The equations of general relativity enable scientists to calculate exactly how much energy is radiated in this manner and thus how quickly the orbits shrink. In 1974 two astrophysicists at the University of Massachusetts Amherst, Joseph Taylor and Russell Hulse, began monitoring a binary neutron star system in the constellation of Aquila using the giant Arecibo radio telescope in Puerto Rico (now decommissioned having fallen into disrepair following hurricane damage). Over several years they were able to measure slight orbital changes and compare their observations with the theoretical prediction. The match was perfect. The result was stunning because it proved that gravitational waves really existed and could transport prodigious amounts of energy across space. It provided a huge fillip to the efforts to detect these elusive waves on Earth.

By this stage the focus had shifted from ringing bars to a completely different method of detection involving very precise laser measurements that would reveal telltale changes in distance due to the stretching and shrinking of space as a gravitational wave passes. Suppose you fire a laser pulse at a distant mirror and time how long it takes to return after reflection. Knowing the speed of light will tell you how far away the mirror is located. Now imagine that a gravitational wave sweeps by and stretches the space between the laser and the mirror. The pulse will get back slightly late. Conversely, if the space is shrunk, the pulse will get back early. That simple notion forms the basis of laser gravitational wave detection.

In practice, timing the pulses individually isn't necessary; a comparison of laser light travelling in different directions is enough. As described in the earlier chapters, the set-up now widely used is the is L-shaped interferometer. Laser beams are sent out perpendicular to each other down each arm of the L and reflected back from distant mirrors. When the light returns, the two reflections are merged. A passing gravitational wave will stretch one arm of the L but shrink the other. If such a change occurs, it can be detected by carefully examining the relative alignments of the peaks and troughs of the light waves from each beam as they are brought together. It is a time-honoured procedure in physics known as wave interference, so the system is known as a laser interferometer.

Part 2 of this book described the two Laser Interferometer Gravitational-Wave Observatory (LIGO) interferometers that were completed in the US in 1999: one at Hanford, Washington, and the other at Livingston, Louisiana. The reason for building two widely separated interferometers was because vibrations can be caused by many sources: earthquakes, waves pounding on sea shores, even freeway traffic or cows. To distinguish between a gravitational wave and, say, a horse galloping in a nearby field, the scientists looked for coincident disturbances in both locations, which could only come from space.

The precision engineering in gravitational wave detectors is mind-boggling. The arms – the sides of each L – are 4-kilometre long high-vacuum tubes. The mirrors suspended at the remote ends are manufactured to extraordinary smoothness and reflectivity. The laser beams are allowed to pass many times back and forth down the tubes before being brought together for comparison, thus increasing the effective arm length. But the really stunning statistics come with the sensitivity attained. Gravitational waves have such a feeble effect that a pulse that was able to shift a mirror by the width of an atom would be unthinkably enormous. So LIGO was designed to detect changes in the mirror's position thousands of times smaller than an atomic nucleus (a 10 trillionth of a centimetre) over a distance of many kilometres. For comparison, it is like being able to detect a change of one hair's breadth in the distance from Earth to the star Alpha Centauri. The sought-after shudders are so tiny that even quantum effects have to be factored in.

Earlier chapters described the nail-biting years when LIGO saw absolutely nothing and it began to look like the instruments might be a white elephant. The scientists went back for more money to do an upgrade and further tweak the sensitivity of the system. Many fingers were crossed. Then, on September 14, 2015, bingo! A distinct judder was detected in both LIGO detectors with identical characteristics.

Furthermore, the disturbance bore all the hallmarks of a burst of gravitational waves emanating from two astronomical bodies spiralling into each other. Calculations soon showed that these objects were too massive to be neutron stars; they had to be black holes, of 36 and 29 solar masses, respectively, located some 1.3 billion light years away. The resulting merged black hole had a mass of sixty-two Suns, so the equivalent of about 3 solar masses had been converted into gravitational wave energy – that's hundreds of times the total energy our Sun has ever emitted as heat during its 4.5-billion-year lifetime, all emitted in a fraction of a second! It was spectacular. All those years of effort had at last paid off. The scientific community was ecstatic. Stephen Hawking told the BBC the detection had 'the potential to revolutionise astronomy'. His sentiments were echoed by then-US President Barack Obama, who tweeted: 'a huge breakthrough in how we understand the universe'.

LIGO finally dispelled any doubts about the reality of gravitational waves, almost a century after Einstein first predicted them. It was a tribute to the tenacity of the scientific community and the confidence they placed in the underlying theory. But the 2015 detection was not the end of the project; rather, it was just the beginning. The ability of laser interferometers to act as super-sensitive ears that can listen to the vibrations of the universe ushered in a brand-new era of astronomy. Since ancient times, astronomers have studied heavenly bodies from the light they emit. Then, in the 1950s, radio telescopes were built, followed by satellites that could detect everything from gamma rays, through X-rays and ultraviolet waves, to infrared rays and microwaves.

Astronomers have now covered the entire electromagnetic spectrum. But gravitational waves are a completely new spectrum, providing a novel window on the universe.

LIGO soon began detecting other binary mergers, and a European system known as Virgo also began operations in Italy. Data from both were used to detect and pinpoint the location of a collision of two neutron stars in August 2017, a remarkable event confirmed when a NASA satellite called Chandra almost simultaneously detected a burst of gamma rays emanating from a mere 130 million light years away in the same patch of sky. Conventional telescopes spotted a rapidly fading luminous source there with the distinct spectral signature of gold and other heavy elements. Astronomers now think impacts between neutron stars have created most of the gold in the universe.

By late 2019, LIGO was running smoothly, and over a period of 4 months it detected no less than thirty-five gravitational wave events – a bonanza that had scientists reaching for the champagne. Of these, most were black hole mergers, a few involved neutron stars, and in one case, a neutron star was observed being swallowed by a black hole. That brought the total number of events to ninety, making it possible to draw some statistical conclusions about the range of black hole masses, their rotation rates and how they formed from the burnt-out cores of massive stars. Since that run, LIGO has been undergoing major upgrades. Scientists hope that with upgrades and the addition of new laser interferometers in India and Japan, gravitational wave events will be observed on a daily basis.

Australia has played a key role in the birth of this new discipline ever since the pioneering work of David Blair on resonant bar detectors. The Australian Research Council funds a centre of excellence for gravitational wave discovery called OzGrav, a consortium involving the University of Western Australia, the Australian National University, Monash University, the University of Melbourne, Swinburne University and the University of Adelaide. In 2020, the Prime Minister's Prize for Science was awarded to the four leaders of the national effort to discover gravitational waves for their critical contributions to the field: David Blair and two other UWA graduates, David McClelland (ANU) and Peter Veitch (Adelaide), along with Victorian Susan Scott (ANU).

Now that gravitational wave astronomy is a reality, scientists are keen to take the next step. The proposed new Australia-based instrument NEMO will focus on gravitational waves of much higher frequency, which will enable astronomers to follow the messy details of neutronic matter sloshing about as pairs of neutron stars scrunch together and gyrate frenetically in the fraction of a second before they plunge down a black hole.

The other end of the spectrum – ultra-low frequency gravitational waves – is not being neglected either. Ambitious plans are afoot to detect them using a giant space-based interferometer that spans 2.5 million kilometres. Known as the Laser Interferometer Space Antenna (LISA), it is being designed by the European Space Agency. The mission will consist of three spacecraft orbiting the Sun in triangular formation, each containing two telescopes, two lasers and two gold-coated test masses. Each spacecraft will be a zero-drag satellite in which the test masses float freely inside a container that shields them from non-gravitational disturbances such as solar wind.

The events detected by LIGO have frequencies in the range of tens to hundreds of hertz (cycles per second) and wavelengths of a few thousand kilometres. The frequency of the peak power radiated by a gravitational source depends on its rate of change. For two stellar-mass black holes spiralling together, each a few kilometres in size, they end up circling at close to the speed of light, which means they orbit each other hundreds of times a second before they merge.

However, black holes come in a variety of sizes. The Milky Way has a large black hole near its centre with a mass of about 4 million suns. Some galaxies are known with still larger black holes containing the mass equivalent of billions of suns. The gravitational waves generated by these objects are much longer and emitted at lower frequency, perhaps taking many minutes to complete just one cycle of the wave. Although collisions between supermassive black holes will be rare, LISA would be sensitive enough to monitor the entire observable universe for such events. More common would be a supermassive black hole swallowing a neutron star or a stellar mass hole, which should also be detectable.

The history of astronomy has shown that each time a new window has been opened on the universe, unexpected discoveries follow. Radio astronomy, developed in the 1950s, led to the accidental discovery of neutron stars in 1967 by a student, Jocelyn Bell, at Cambridge University. Bell was alerted to something unusual: highly regular radio pulses, which turned out to be the signature of spinning neutron stars (and is the reason these objects are referred to as pulsars). The launch of X-ray satellites in the early 1970s led to the first identification of black holes, after it was found that when a black hole is in orbit around a normal star, it drags away and swallows some of the stellar material, heating it so much on the way that it emits X-rays. It is a similar story for infrared-, ultraviolet- and gamma-ray orbiting observatories.

Few doubt that the same will prove true of the new gravitational window. Because they interact so weakly with matter, gravitational waves can emanate from regions that are shrouded from the view of optical telescopes and other electromagnetic observations. To date, most of the action has centred around black holes and neutron stars, but any movement of large masses will generate gravitational waves. Theoretical astrophysicists have a long list of exotic hypothetical cosmic objects, any of which could be copious sources of gravitational waves displaying distinctive features. For example, some astrophysicists believe that as the universe cooled from the heat of the Big Bang, thread-like entities formed, concentrating enormous mass into infinitesimally thin tubes. These hypothetical 'cosmic strings' would be strong sources of gravitational waves as they thrash about. And the most violent event in cosmic history was the Big Bang itself – the explosion that marked the birth of the whole universe. It would have filled the cosmos with gravitational waves that are rumbling through space still. If these primordial disturbances could be detected, they would provide vital clues to the earliest epoch of physical existence and the forces that shaped the universe we see today.

*there are certain to be phenomena
that nobody has yet thought of*

However fertile the imagination of theoretical astrophysicists, there are certain to be phenomena that nobody has yet thought of. Because every physical process is a source of gravitational waves, the promise of gravitational astronomy is that it can, in time, help compile the most comprehensive catalogue of all the objects and systems out there in the million trillion trillion cubic light years of space that make up the observable universe. Armed with this inventory of astronomical entities, scientists will be able to reconstruct the complete narrative of the birth, evolution and likely death of the universe. The obscure vibrations of space predicted mathematically by Einstein a century ago serve as a key to unlocking the secrets of the universe, and this generation will go down in history as the one that began that quest.

ACKNOWLEDGEMENTS

This book would not have been possible without the support of the Einstein-First team whose desire to share our best understanding of the physical universe has created a stimulating environment where we all challenge each other to find better ways of explaining the revelations of modern science at a level where everyone can understand. David Wood, Jyoti Kaur, David Treagust and PhD students Rahul Choudhery, Shon Boublil, Kyla Adams and Anastasia Popkova have all contributed. Jesse Santoso has contributed their passion for education combined with video production skills to helping us train teachers in the modern concepts of physical reality. Special thanks go to Peter Rossdeutscher and Howard Golden as well as Elaine Horne and Marjan Zadnik whose support has helped our whole team.

I wish to thank my OzGrav colleagues who have created the ongoing climate of discovery and enquiry which will take us into the future, and which provides the platform of deep understanding needed for translating research discoveries into public understanding. These people include Jackie Bondell and Matthew Bailes from Swinburne University, Susan Scott from ANU, as well as the team of younger researchers who are planning the future NEMO observatory for Australia.

I am especially indebted to our financial sponsors who have shared our vision of bringing Einstein-First, and with it our modern understanding of the Universe, to schools all over Australia. We are also enormously grateful to the many teachers who have embraced our programs with enthusiasm and taught us how scary and bewildering it can be to change your thinking from the old paradigm to the new one based on Einsteinian science.

ACKNOWLEDGEMENTS

I want to thank Ju Li, my wife, and Carl Blair, my son, who have contributed to the book and to my other children Palenque, Linden and Julian, as well as Damon, Felix and Hanna who I thank for their help, their questions, their tolerance and their enthusiasm.

I want to thank my co-authors. Ron Burman did most of the historical research and fed me continuously with more and more historical material that enriched the story of Wallal. Paul Davies, who wrote the very first popular book on gravitational wave detection, contributed his deep understanding to tell the story of the near future dreams of gravitational wave physicists.

The book tells briefly about the mammoth struggle of building the first gravitational wave detector Niobe, in Australia. This was done by a team of dedicated PhD students that included Tony Mann, John Ferreirinho, Peter Veitch, Peter Turner, Nick Linthorne, Laurie Mann, Mike Tobar, Ik Siong Heng, Steve Jones and many others, with lots of support from Cyril Edwards. Frank Van Kann and Michael Buckingham.

I want to thank especially Ruby Chan for her invaluable support, my editor Sam Trafford and Kate Pickard, the publisher, who brought the book together.

Below we list all the organisations who have supported Einstein-First and made it possible to create this book as a component of our Einstein-First initiative.

David Blair
August 2022

Ingram Content Group Australia Pty Ltd
Printed in Australia
AUHW011012200623
379741AU00001B/1